STATISTICS: Textbooks and Monographs

D. B. Owen

Founding Editor, 1972–1991

Associate Editors

Statistical Computing/ Nonparametric Statistics
Professor William R. Schucany
Southern Methodist University

Probability
Professor Marcel F. Neuts
University of Arizona

Multivariate Analysis
Professor Anant M. Kshirsagar
University of Michigan

Quality Control/Reliability
Professor Edward G. Schilling
Rochester Institute of Technology

Editorial Board

Applied Probability
Dr. Paul R. Garvey
The MITRE Corporation

Economic Statistics
Professor David E. A. Giles
University of Victoria

Experimental Designs
Mr. Thomas B. Barker
Rochester Institute of Technology

Multivariate Analysis
Professor Subir Ghosh
University of California, Riverside

Statistical Distributions
Professor N. Balakrishnan
McMaster University

Statistical Process Improvement
Professor G. Geoffrey Vining
Virginia Polytechnic Institute

Stochastic Processes
Professor V. Lakshmikantham
Florida Institute of Technology

Survey Sampling
Professor Lynne Stokes
Southern Methodist University

Time Series
Sastry G. Pantula
North Carolina State University

STATISTICS: Textbooks and Monographs

Recent Titles

Statistics for the 21st Century: Methodologies for Applications of the Future, *edited by C. R. Rao and Gábor J. Székely*

Probability and Statistical Inference, *Nitis Mukhopadhyay*

Handbook of Stochastic Analysis and Applications, *edited by D. Kannan and V. Lakshmikantham*

Testing for Normality, *Henry C. Thode, Jr.*

Handbook of Applied Econometrics and Statistical Inference, *edited by Aman Ullah, Alan T. K. Wan, and Anoop Chaturvedi*

Visualizing Statistical Models and Concepts, *R. W. Farebrother and Michaël Schyns*

Financial and Actuarial Statistics: An Introduction, *Dale S. Borowiak*

Nonparametric Statistical Inference, Fourth Edition, Revised and Expanded, *Jean Dickinson Gibbons and Subhabrata Chakraborti*

Computer-Aided Econometrics, *edited by David E.A. Giles*

The EM Algorithm and Related Statistical Models, *edited by Michiko Watanabe and Kazunori Yamaguchi*

Multivariate Statistical Analysis, Second Edition, Revised and Expanded, *Narayan C. Giri*

Computational Methods in Statistics and Econometrics, *Hisashi Tanizaki*

Applied Sequential Methodologies: Real-World Examples with Data Analysis, *edited by Nitis Mukhopadhyay, Sujay Datta, and Saibal Chattopadhyay*

Handbook of Beta Distribution and Its Applications, *edited by Arjun K. Gupta and Saralees Nadarajah*

Item Response Theory: Parameter Estimation Techniques, Second Edition, *edited by Frank B. Baker and Seock-Ho Kim*

Statistical Methods in Computer Security, *edited by William W. S. Chen*

Elementary Statistical Quality Control, Second Edition, *John T. Burr*

Data Analysis of Asymmetric Structures, *Takayuki Saito and Hiroshi Yadohisa*

Mathematical Statistics with Applications, *Asha Seth Kapadia, Wenyaw Chan, and Lemuel Moyé*

Advances on Models, Characterizations and Applications, *N. Balakrishnan, I. G. Bairamov, and O. L. Gebizlioglu*

Survey Sampling: Theory and Methods, Second Edition, *Arijit Chaudhuri and Horst Stenger*

Statistical Design of Experiments with Engineering Applications, *Kamel Rekab and Muzaffar Shaikh*

Quality by Experimental Design, Third Edition, *Thomas B. Barker*

Handbook of Parallel Computing and Statistics, *Erricos John Kontoghiorghes*

Statistical Inference Based on Divergence Measures, *Leandro Pardo*

A Kalman Filter Primer, *Randy Eubank*

Introductory Statistical Inference, *Nitis Mukhopadhyay*

Handbook of Statistical Distributions with Applications, *K. Krishnamoorthy*

Handbook of Statistical Distributions with Applications

K. Krishnamoorthy
University of Louisiana at Lafayette
U.S.A.

Published in 2006 by
Chapman & Hall/CRC Press
Taylor & Francis Group
6000 Broken Sound Parkway NW, Suite 300
Boca Raton, FL 33487-2742

© 2006 by Taylor & Francis Group, LLC
Chapman & Hall/CRC Press is an imprint of Taylor & Francis Group

No claim to original U.S. Government works
Printed in the United States of America on acid-free paper
10 9 8 7 6 5 4 3 2 1

International Standard Book Number-10: 1-5848-8635-8 (Hardcover)
International Standard Book Number-13: 978-1-5848-8635-8 (Hardcover)

This book contains information obtained from authentic and highly regarded sources. Reprinted material is quoted with permission, and sources are indicated. A wide variety of references are listed. Reasonable efforts have been made to publish reliable data and information, but the author and the publisher cannot assume responsibility for the validity of all materials or for the consequences of their use.

No part of this book may be reprinted, reproduced, transmitted, or utilized in any form by any electronic, mechanical, or other means, now known or hereafter invented, including photocopying, microfilming, and recording, or in any information storage or retrieval system, without written permission from the publishers.

For permission to photocopy or use material electronically from this work, please access www.copyright.com (http://www.copyright.com/) or contact the Copyright Clearance Center, Inc. (CCC) 222 Rosewood Drive, Danvers, MA 01923, 978-750-8400. CCC is a not-for-profit organization that provides licenses and registration for a variety of users. For organizations that have been granted a photocopy license by the CCC, a separate system of payment has been arranged.

Trademark Notice: Product or corporate names may be trademarks or registered trademarks, and are used only for identification and explanation without intent to infringe.

Library of Congress Cataloging-in-Publication Data

Krishnamoorthy, K. (Kalimuthu)
Handbook of statistical distributions with applications / K. Krishnamoorthy.
p. cm. -- (Statistics, a series of textbooks & monographs ; 188)
Includes bibliographical references and index.
ISBN 1-58488-635-8
1. Distribution (probability theory)--Handbooks, manuals, etc. I. Title. II. Series: Statistics, textbooks and monographs ; v. 188.

QA273.6.K75 2006
519.5--dc22 2006040297

Visit the Taylor & Francis Web site at
http://www.taylorandfrancis.com

and the CRC Press Web site at
http://www.crcpress.com

In memory of my parents

Preface

Statistical distributions and models are commonly used in many applied areas such as economics, engineering, social, health, and biological sciences. In this era of inexpensive and faster personal computers, practitioners of statistics and scientists in various disciplines have no difficulty in fitting a probability model to describe the distribution of a real-life data set. Indeed, statistical distributions are used to model a wide range of practical problems, from modeling the size grade distribution of onions to modeling global positioning data. Successful applications of these probability models require a thorough understanding of the theory and familiarity with the practical situations where some distributions can be postulated. Although there are many statistical software packages available to fit a probability distribution model for a given data set, none of the packages is comprehensive enough to provide table values and other formulas for numerous probability distributions. The main purpose of this book and the software is to provide users with quick and easy access to table values, important formulas, and results of the many commonly used, as well as some specialized, statistical distributions. The book and the software are intended to serve as reference materials. With practitioners and researchers in disciplines other than statistics in mind, I have adopted a format intended to make it simple to use the book for reference purposes. Examples are provided mainly for this purpose.

I refer to the software that computes the table values, moments, and other statistics as *StatCalc*. For rapid access and convenience, many results, formulas and properties are provided for each distribution. Examples are provided to illustrate the applications of *StatCalc*. The *StatCalc* is a dialog-based application, and it can be executed along with other applications.

The programs of *StatCalc* are coded in C++ and compiled using Microsoft Visual C++ 6.0. All intermediate values are computed using double precision so that the end results will be more accurate. I compared the table values of *StatCalc* with the classical hard copy tables such as *Biometrika Tables for Statisticians*, *Hand-book of Mathematical Functions* by Abramowitz and Stegun (1965), *Tables of the Bivariate Normal Distribution Function and Related Functions* by National Bureau of Standards 1959, *Pocket Book of Statistical*

Tables by Odeh, et. al. (1977), and the tables published in various journals listed in the references. Table values of the distributions of Wilcoxon Rank-Sum Statistic and Wilcoxon Signed-Rank Statistic are compared with those given in *Selected Tables in Mathematical Statistics.* The results are in agreement wherever I checked. I have also verified many formulas and results given in the book either numerically or analytically. All algorithms for random number generation and evaluating cumulative distribution functions are coded in Fortran, and verified for their accuracy. Typically, I used 1,000,000 iterations to evaluate the performance of random number generators in terms of the speed and accuracy. All the algorithms produced satisfactory results. In order to avoid typographical errors, algorithms are created by copying and editing the Fortran codes used for verification.

A reference book of this nature cannot be written without help from numerous people. I am indebted to many researchers who have developed the results and algorithms given in the book. I would like to thank my colleagues for their valuable help and suggestions. Special thanks are due to Tom Rizzuto for providing me numerous books, articles, and journals. I am grateful to computer science graduate student Prasad Braduleker for his technical help at the initial stage of the *StatCalc* project. It is a pleasure to thank P. Vellaisamy at IIT – Bombay who thoroughly read and commented on the first fourteen chapters of the book. I am thankful to my graduate student Yanping Xia for checking the formulas and the software *StatCalc* for accuracies.

K. Krishnamoorthy
University of Louisiana at Lafayette

Contents

INTRODUCTION TO STATCALC

1 PRELIMINARIES

2 DISCRETE UNIFORM DISTRIBUTION

3 BINOMIAL DISTRIBUTION

4 HYPERGEOMETRIC DISTRIBUTION

5 POISSON DISTRIBUTION

6 GEOMETRIC DISTRIBUTION

7 NEGATIVE BINOMIAL DISTRIBUTION

8 LOGARITHMIC SERIES DISTRIBUTION

9 UNIFORM DISTRIBUTION

10 NORMAL DISTRIBUTION

11 CHI-SQUARE DISTRIBUTION

12 F DISTRIBUTION

13 STUDENT'S t DISTRIBUTION

14 EXPONENTIAL DISTRIBUTION

15 GAMMA DISTRIBUTION

16 BETA DISTRIBUTION

17 NONCENTRAL CHI-SQUARE DISTRIBUTION

18 NONCENTRAL F DISTRIBUTION

19 NONCENTRAL t DISTRIBUTION

20 LAPLACE DISTRIBUTION

21 LOGISTIC DISTRIBUTION

22 LOGNORMAL DISTRIBUTION

23 PARETO DISTRIBUTION

24 WEIBULL DISTRIBUTION

25 EXTREME VALUE DISTRIBUTION

26 CAUCHY DISTRIBUTION

27 INVERSE GAUSSIAN DISTRIBUTION

28 RAYLEIGH DISTRIBUTION

29 BIVARIATE NORMAL DISTRIBUTION

30 DISTRIBUTION OF RUNS

31 SIGN TEST AND CONFIDENCE INTERVAL FOR THE MEDIAN

32 WILCOXON SIGNED-RANK TEST

33 WILCOXON RANK-SUM TEST

34 NONPARAMETRIC TOLERANCE INTERVAL

35 TOLERANCE FACTORS FOR A MULTIVARIATE NORMAL POPULATION

36 DISTRIBUTION OF THE SAMPLE MULTIPLE CORRELATION COEFFICIENT

Introduction to StatCalc

0.1 Introduction

The software accompanying this book is referred to as *StatCalc*, which is a PC calculator that computes various statistical table values. More specifically, it computes table values of all the distributions presented in the book, necessary statistics to carry out some hypothesis tests and to construct confidence intervals, required sample sizes to carry out a test within the specified accuracies, and much more. Readers who are familiar with some statistical concepts and terminologies, and PC calculators may find *StatCalc* as simple and easy to use. In the following, we explain how to use this program and illustrate some features.

The dialog boxes that compute various table values are grouped into 4 categories, namely, continuous, discrete, nonparametric and miscellaneous as shown in the main page of *StatCalc* in Figure 0.1(a). Let us assume we want to compute binomial probabilities; if so then we should first select "Discrete dialog box" (by clicking on the radio button [Discrete]) as the binomial distribution is a discrete distribution (see Figure 0.1(b)). Click on [Binomial], and then click on [Probabilities, Critical Values and Moments] to get the binomial probability dialog box. This sequence of selections is indicated in the book by the trajectory [StatCalc→Discrete→Binomial→Probabilities, Critical Values and Moments]. Similarly, if we need to compute factors for constructing tolerance intervals for a normal distribution, we first select [Continuous] (because the normal distribution is a continuous one), and then select [Normal] and [Tolerance Limits]. This sequence of selections is indicated by the trajectory [StatCalc→Continuous→Normal→Tolerance Limits]. After selecting the desired dialog box, input the parameters and other values to compute the needed table values.

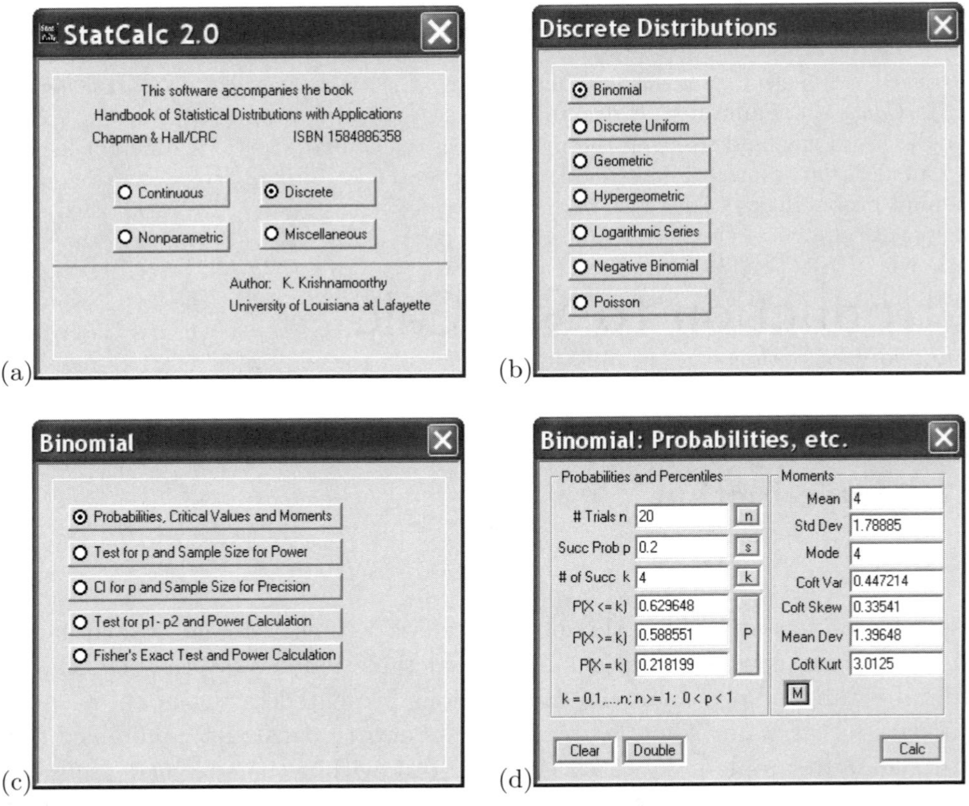

Figure 0.1 Selecting the Dialog Box for Computing Binomial Probabilities

StatCalc is a stand alone application, and many copies (as much as the screen can hold) of *StatCalc* can be opened simultaneously. To open two copies, click on *StatCalc* icon on your desktop or select from the start menu. Once the main page of *StatCalc* opens, click on *StatCalc* icon again on your desktop. The second copy of *StatCalc* pops up exactly over the first copy, and so using the mouse drag the second copy to a different location on your desktop. Now, we have two copies of *StatCalc*. Suppose we want to compare binomial probabilities with those of the hypergeometric with lot size 5000, then select binomial from one of the copies and hypergeometric from the other. Input the values as shown in Figure 0.2. We observe from these two dialog boxes that the binomial probabilities with $n = 20$ and $p = 0.2$ are very close to those of the hypergeometric with lot size (population size) 5000. Furthermore, good agreement of the moments of these two distributions clearly indicates that, when the lot size is 5000 or more, the hypergeometric probabilities can be safely approximated by the binomial probabilities.

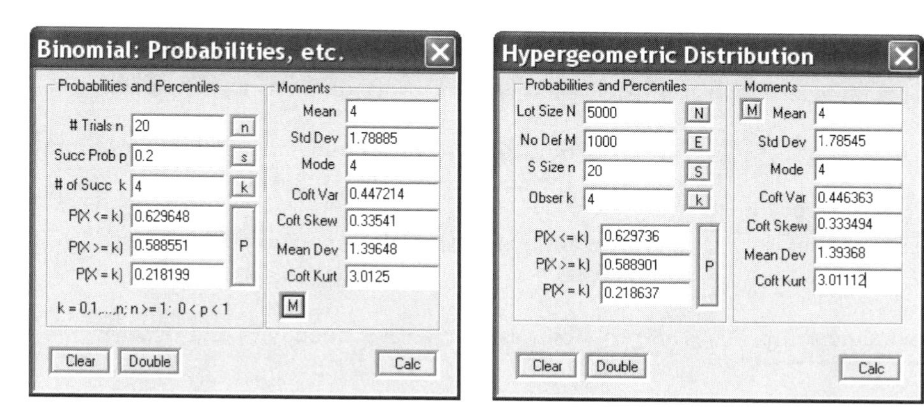

Figure 0.2 Dialog Boxes for Computing Binomial and Hypergeometric Probabilities

StatCalc can be opened along with other applications, and the values from the edit boxes (the white boxes) can be copied [Ctrl+c] and pasted [Ctrl+v] in a document.

0.2 Contents of StatCalc

Continuous Distributions	
1 Beta	Tail probabilities, percentiles, moments and other parameters.
2 Bivariate Normal	All tail probabilities; test and confidence interval for the correlation coefficient; test and confidence interval for the difference between two independent correlation coefficients.
3 Cauchy	Tail probabilities, percentiles and other parameters.
4 Chi-square	Tail probabilities, percentiles and moments; also computes degrees of freedom when other values are given.
5 Exponential	Tail probabilities, percentiles, moments and other parameters.
6 Extreme Value	Tail probabilities, percentiles, moments and other parameters.
7 F Distribution	Tail probabilities, percentiles, moments; also computes the degrees of freedoms when other values are given.
8 Gamma	Tail probabilities, percentiles, moments and other parameters; Test and confidence interval for the scale parameter.
9 Inverse Gaussian	Tail probabilities, percentiles, moments and other parameters; test and confidence interval for the mean; test and confidence interval for the difference between two means; test and confidence interval for the ratio of two means.
10 Laplace	Tail probabilities, percentiles, moments and other parameters.
11 Logistic	Tail probabilities, percentiles, moments and other parameters.

12 Lognormal	Tail probabilities, percentiles, moments and other parameters; t-test and confidence interval for the mean; test and confidence interval for the difference between two means; test and confidence interval for the ratio of two means.
13 Noncentral χ^2	Tail probabilities, percentiles and moments; computation of the degrees of freedom and noncentrality parameter.
14 Noncentral F	Tail probabilities, percentiles and moments; calculation of the degrees of freedom and noncentrality parameter.
15 Noncentral t	Tail probabilities, percentiles and moments; computation of the degrees of freedom and noncentrality parameter.
16 Normal	Tail probabilities, percentiles, and moments; test and confidence interval for the mean; power of the t-test; test and confidence interval for the variance; test and confidence interval for the variance ratio; two-sample t-test and confidence interval; two-sample test with no assumption about the variances; power of the two-sample t-test; tolerance intervals for a normal distribution; tolerance intervals controlling both tails; simultaneous tests for quantiles; tolerance limits for one-way random effects model.
17 Pareto	Tail probabilities, percentiles, moments, and other parameters.
18 Rayleigh	Tail probabilities, percentiles, moments, and other parameters.
19 Student's t	Tail probabilities, percentiles, and moments; also computes the degrees of freedom when other values are given; computes tail probabilities and critical values of the distribution of the maximum of several $\|t\|$ variables.
20 Weibull	Tail probabilities, percentiles, moments and other parameters.

Discrete Distributions	
21 Binomial	Tail probabilities, critical values, moments, and other parameters; test for the proportion and power calculation; confidence intervals for the proportion and sample size for precision; test for the difference between two proportions and power calculation; Fisher's exact test and power calculation.
22 Discrete Uniform	Tail probabilities and moments.
23 Geometric	Tail probabilities, critical values, and moments; confidence interval for success probability;
24 Hypergeometric	Tail probabilities, critical values and moments; test for the proportion and power calculation; confidence interval and sample size for precision; test for the difference between proportions and power calculation.
25 Logarithmic Series	Tail probabilities, critical values and moments.
26 Negative Binomial	Tail probabilities, critical values, and moments; test for the proportion and power calculation; confidence intervals for the proportion.
27 Poisson	Tail probabilities, critical values and moments; test for the mean and power calculation; confidence interval for mean and sample size for precision; test for the ratio of two means, and power calculation; confidence intervals for the ratio of two means; test for the difference between two means and power calculation.

Nonparametric	
28 Distribution of Runs	Tail probabilities and critical values.
29 Sign Test and Confidence Interval for the Median	Nonparametric test for the median; also computes confidence intervals for the median.
30 Wilcoxon signed-rank test	Computes the p-values and critical values for testing the median.
31 Wilcoxon rank-sum test	Computes p-values for testing equality of two distributions; Moments and critical values.
32 Nonparametric tolerance limits	Computes size of the sample so that the smallest and the largest order statistics form a tolerance interval.
Miscellaneous	
33 Tolerance factors for a multivariate normal population	Computes factors for constructing tolerance region for a multivariate normal population.
34 Distribution of the sample multiple correlation coefficient	Test and confidence interval for the squared multiple correlation coefficient.

Chapter 1

Preliminaries

This reference book is written for those who have some knowledge of statistical distributions. In this chapter we will review some basic terms and concepts, and introduce the notations used in the book. Readers should be familiar with these concepts in order to understand the results, formulas, and properties of the distributions presented in the rest of the book. This chapter also covers two standard methods of fitting a distribution for an observed data set, two classical methods of estimation, and some aspects of hypothesis testing and interval estimation. Furthermore, some methods for generating random numbers from a probability distribution are outlined.

1.1 Random Variables and Expectations

Random Experiment: An experiment whose outcomes are determined only by chance factors is called a random experiment.

Sample Space: The set of all possible outcomes of a random experiment is called a sample space.

Event: The collection of none, one, or more than one outcomes from a sample space is called an event.

Random Variable: A variable whose numerical values are determined by chance factors is called a random variable. Formally, it is a function from the sample space to a set of real numbers.

Discrete Random Variable: If the set of all possible values of a random variable X is countable, then X is called a discrete random variable.

Probability of an Event: If all the outcomes of a random experiment are equally likely, then the probability of an event A is given by

$$P(A) = \frac{\text{Number of outcomes in the event A}}{\text{Total number of outcomes in the sample space}}.$$

Probability Mass Function (pmf): Let R be the set of all possible values of a discrete random variable X, and $f(k) = P(X = k)$ for each k in R. Then $f(k)$ is called the probability mass function of X. The expression $P(X = k)$ means the probability that X assumes the value k.

Example 1.1 A fair coin is to be flipped three times. Let X denote the number of heads that can be observed out of these three flips. Then X is a discrete random variable with the set of possible values {0, 1, 2, 3}; this set is also called the *support* of X. The sample space for this example consists of all possible outcomes ($2^3 = 8$ outcomes) that could result out of three flips of a coin, and is given by

$$\{HHH, HHT, HTH, THH, HTT, THT, TTH, TTT\}.$$

Note that all the above outcomes are equally likely to occur with a chance of 1/8. Let A denote the event of observing two heads. The event A occurs if one of the outcomes HHT, HTH, and THH occurs. Therefore, $P(A) = 3/8$. The probability distribution of X can be obtained similarly and is given below:

k:	0	1	2	3
$P(X = k)$:	1/8	3/8	3/8	1/8

This probability distribution can also be obtained using the probability mass function. For this example, the pmf is given by

$$P(X = k) = \binom{3}{k}\left(\frac{1}{2}\right)^k\left(1 - \frac{1}{2}\right)^{3-k}, \quad k = 0, 1, 2, 3,$$

and is known as the binomial$(3, 1/2)$ mass function (see Chapter 3).

Continuous Random Variable: If the set of all possible values of X is an interval or union of two or more nonoverlapping intervals, then X is called a continuous random variable.

Probability Density Function (pdf): Any real valued function $f(x)$ that satisfies the following requirements is called a probability density function:

$$f(x) \geq 0 \text{ for all } x, \text{ and } \int_{-\infty}^{\infty} f(x)dx = 1.$$

Cumulative Distribution Function (cdf): The cdf of a random variable X is defined by

$$F(x) = P(X \leq x)$$

for all x. For a continuous random variable X with the probability density function $f(x)$,

$$P(X \leq x) = \int_{-\infty}^{x} f(t)dt \text{ for all } x.$$

For a discrete random variable X, the cdf is defined by

$$F(k) = P(X \leq k) = \sum_{i=-\infty}^{k} P(X = i).$$

Many commonly used distributions involve constants known as parameters. If the distribution of a random variable X depends on a parameter θ (θ could be a vector), then the pdf or pmf of X is usually expressed as $f(x|\theta)$, and the cdf is written as $F(x|\theta)$.

Inverse Distribution Function: Let X be a random variable with the cdf $F(x)$. For a given $0 < p < 1$, the inverse of the distribution function is defined by

$$F^{-1}(p) = \inf\{x : P(X \leq x) = p\}.$$

Expectation: If X is a continuous random variable with the pdf $f(x)$, then the expectation of $g(X)$, where g is a real valued function, is defined by

$$E(g(X)) = \int_{-\infty}^{\infty} g(x)f(x)dx.$$

If X is a discrete random variable, then

$$E(g(X)) = \sum_{k} g(k)P(X = k),$$

where the sum is over all possible values of X. Thus, $E(g(X))$ is the weighted average of the possible values of $g(X)$, each weighted by its probability.

1.2 Moments and Other Functions

The moments are a set of constants that represent some important properties of the distributions. The most commonly used such constants are measures of central tendency (mean, median, and mode), and measures of dispersion (variance and mean deviation). Two other important measures are the coefficient of skewness and the coefficient of kurtosis. The coefficient of skewness measures the degree of asymmetry of the distribution, whereas the coefficient of kurtosis measures the degree of flatness of the distribution.

1.2.1 Measures of Central Tendency

Mean: Expectation of a random variable X is called the mean of X or the mean of the distribution of X. It is a measure of location of all possible values of X. The mean of a random variable X is usually denoted by μ, and for a discrete random variable X it is defined by

$$\mu = E(X) = \sum_k kP(X = k),$$

where the sum is over all possible values of X. For a continuous random variable X with probability density function $f(x)$, the mean is defined by

$$\mu = E(X) = \int_{-\infty}^{\infty} xf(x)dx.$$

Median: The median of a continuous random variable X is the value such that 50% of the possible values of X are less than or equal to that value. For a discrete distribution, median is not well defined, and it need not be unique (see Example 1.1).

Mode: The most probable value of the random variable is called the mode.

1.2.2 Moments

Moments about the Origin (Raw Moments): The moments about the origin are obtained by finding the expected value of the random variable that has been raised to k, $k = 1, 2, \ldots$. That is,

$$\mu_k' = E(X^k) = \int_{-\infty}^{\infty} x^k f(x)dx$$

is called the kth moment about the origin or the kth raw moment of X.

Moments about the Mean (Central Moments): When the random variable is observed in terms of deviations from its mean, its expectation yields moments about the mean or central moments. The first central moment is zero, and the second central moment is the variance. The third central moment measures the degree of skewness of the distribution, and the fourth central moment measures the degree of flatness. The kth moment about the mean or the kth central moment of a random variable X is defined by

$$\mu_k = E(X - \mu)^k, \quad k = 1, 2, \ldots,$$

where $\mu = E(X)$ is the mean of X. Note that the first central moment μ_1 is always zero.

Sample Moments: The sample central moments and raw moments are defined analogous to the moments defined above. Let $X_1, \ldots, X_n$ be a sample from a population. The sample kth moment about the origin is defined by

$$m'_k = \frac{1}{n}\sum_{i=1}^{n} X_i^k, \quad k = 1, 2, \ldots$$

and the sample kth moment about the mean is defined by

$$m_k = \frac{1}{n}\sum_{i=1}^{n} (X_i - \bar{X})^k, \quad k = 1, 2, \ldots,$$

where $\bar{X} = m'_1$. In general, for a real valued function g, the sample version of $E(g(X))$ is given by $\sum_{i=1}^{n} g(X_i)/n$.

1.2.3 Measures of Variability

Variance: The second moment about the mean (or the second central moment) of a random variable X is called the variance and is usually denoted by σ^2. It is a measure of the variability of all possible values of X. The positive square root of the variance is called the *standard deviation*.

Coefficient of Variation: Coefficient of variation is the ratio of the standard deviation and the mean, that is, (σ/μ). This is a measure of variability independent of the scale. That is, coefficient of variation is not affected by the units of measurement. Note that the variance is affected by the units of measurement.

Mean Deviation: Mean deviation is a measure of variability of the possible values of the random variable X. It is defined as the expectation of absolute difference between X and its mean. That is,

$$\text{Mean Deviation} = E(|X - \mu|).$$

1.2.4 Measures of Relative Standing

Percentile (quantile): For a given $0 < p < 1$, the $100p$th percentile of a distribution function $F(x)$ is the value of x for which $F(x) = p$. That is, $100p\%$ of the population data are less than or equal to x. If a set of values of x satisfy $F(x) = p$, then the minimum of the set is the $100p$th percentile. The $100p$th percentile is also called the pth quantile.

Quartiles: The 25th and 75th percentiles are, respectively, called the first and the third quartile. The difference (third quartile – first quartile) is called the *inter quartile range*.

1.2.5 Other Measures

Coefficient of Skewness: The coefficient of skewness is a measure of skewness of the distribution of X. If the coefficient of skewness is positive, then the distribution is skewed to the right; that is, the distribution has a long right tail. If it is negative, then the distribution is skewed to the left. The coefficient of skewness is defined as

$$\frac{\text{Third Moment about the Mean}}{(\text{Variance})^{\frac{3}{2}}} = \frac{\mu_3}{\mu_2^{\frac{3}{2}}}$$

Coefficient of Kurtosis:

$$\gamma_2 = \frac{\text{4th Moment about the Mean}}{(\text{Variance})^2} = \frac{\mu_4}{\mu_2^2}$$

is called the coefficient of kurtosis or coefficient of excess. This is a scale and location invariant measure of degree of *peakedness* of the probability density curve. If $\gamma_2 < 3$, then the probability density curve is called *platykurtic*; if $\gamma_2 > 3$, it is called *lepto kurtic*; if $\gamma_2 = 3$, it is called *mesokurtic*.

Coefficient of skewness and coefficient of kurtosis are useful to approximate the distribution of X. For instance, if the distribution of a random variable Y is known, and its coefficient of skewness and coefficient of kurtosis are approximately equal to those of X, then the distribution functions of X and Y are approximately equal. In other words, X is approximately distributed as Y.

1.2.6 Some Other Functions

Moment Generating Function: The moment generating function of a random variable X is defined by

$$M_X(t) = E\left(e^{tX}\right),$$

provided that the expectation exists for t in some neighborhood of zero. If the expectation does not exist for t in a neighborhood of zero, then the moment generating function does not exist. The moment generating function is useful in deriving the moments of X. Specifically,

$$E(X^k) = \left.\frac{\partial^k E(e^{tx})}{\partial t^k}\right|_{t=0}, \quad k = 1, 2, \ldots$$

Characteristic Function: The characteristic function of a random variable X is defined by

$$\phi_X(t) = E\left(e^{itX}\right),$$

where i is the complex number and t is a real number. Every random variable has a unique characteristic function. Therefore, the characteristic function of X uniquely determines its distribution.

Probability Generating Function: The probability generating function of a nonnegative, integer valued random variable X is defined by

$$P(t) = \sum_{i=0}^{\infty} t^i P(X = i)$$

so that

$$P(X = k) = \frac{1}{k!}\left.\left(\frac{d^k P(t)}{dt^k}\right)\right|_{t=0}, \quad k = 1, 2, \ldots$$

Furthermore, $P(0) = P(X = 0)$ and $\left.\frac{dP(t)}{dt}\right|_{t=1} = E(X)$.

1.3 Some Functions Relevant to Reliability

Survival Function: The survival function of a random variable X with the distribution function $F(x)$ is defined by

$$1 - F(x) = P(X > x).$$

If X represents the life of a component, then the value of the survival function at x is called the survival probability (or reliability) of the component at x.

Inverse Survival Function: For a given probability p, the inverse survival function returns the value of x that satisfies $P(X > x) = p$.

Hazard Rate: The hazard rate of a random variable at time x is defined by

$$r(x) = \frac{f(x)}{1 - F(x)}.$$

Hazard rate is also referred to as failure rate, intensity rate, and force of mortality. The survival probability at x in terms of the hazard rate is given by

$$P(X > x) = \exp\left(-\int_0^x r(y)dy\right).$$

Hazard Function: The cumulative hazard rate

$$R(x) = \int_0^x \frac{f(y)}{1 - F(y)}dy$$

is called the hazard function.

Increasing Failure Rate (IFR): A distribution function $F(x)$ is said to have increasing failure rate if

$$P(X > x|t) = \frac{P(X > t + x)}{P(X > t)} \text{ is decreasing in time } t \text{ for each } x > 0.$$

Decreasing Failure Rate (DFR): A distribution function $F(x)$ is said to have decreasing failure rate if

$$P(X > x|t) = \frac{P(X > t + x)}{P(X > t)} \text{ is increasing in time } t \text{ for each } x > 0.$$

1.4 Model Fitting

Let $X_1, \ldots, X_n$ be a sample from a continuous population. To verify whether the data can be modeled by a continuous distribution function $F(x|\theta)$, where θ is an unknown parameter, the plot called Q–Q plot can be used. If the sample size is 20 or more, the Q–Q plot can be safely used to check whether the data fit the distribution.

1.4.1 Q–Q Plot

Construction of a Q–Q plot involves the following steps:

1. Order the sample data in ascending order and denote the jth smallest observation by $x_{(j)}$, $j = 1, \ldots, n$. The $x_{(j)}$'s are called *order statistics* or sample quantiles.
2. The proportion of data less than or equal to $x_{(j)}$ is usually approximated by $(j - 1/2)/n$ for theoretical convenience.
3. Find an estimator $\widehat{\theta}$ of θ (θ could be a vector).
4. Estimate the population quantile $q_{(j)}$ as the solution of the equation

$$F(q_{(j)}|\widehat{\theta}) = (j - 1/2)/n, \quad j = 1, \ldots, n.$$

5. Plot the pairs $(x_{(1)}, q_{(1)}), \ldots, (x_{(n)}, q_{(n)})$.

If the sample is from a population with the distribution function $F(x|\theta)$, then the Q–Q plot forms a line pattern close to the $y = x$ line, because the sample quantiles and the corresponding population quantiles are expected to be equal. If this happens, then the distribution model $F(x|\theta)$ is appropriate for the data (for examples, see Sections 10.1 and 16.5).

The following chi-square goodness-of-fit test may be used if the sample is large or the data are from a discrete population.

1.4.2 The Chi-Square Goodness-of-Fit Test

Let X be a discrete random variable with the support $\{x_1, ..., x_m\}$. Assume that $x_1 \leq ... \leq x_m$. Let $X_1, \ldots, X_n$ be a sample of n observations on X. Suppose we hypothesize that the sample is from a particular discrete distribution with the probability mass function $f(k|\theta)$, where θ is an unknown parameter (it could be a vector). The hypothesis can be tested as follows.

1. Find the number O_j of data points that are equal to x_j, $j = 1, 2, \ldots, m$. The O_j's are called observed frequencies.
2. Compute an estimator $\widehat{\theta}$ of θ based on the sample.
3. Compute the probabilities $p_j = f(x_j|\widehat{\theta})$ for $j = 1, 2, \ldots, m - 1$ and $p_m = 1 - \sum_{j=1}^{m-1} p_j$.

4. Compute the expected frequencies $E_j = p_j \times n, \quad j = 1, \ldots, m$.

5. Evaluate the chi-square statistic

$$\chi^2 = \sum_{j=1}^{m} \frac{(O_j - E_j)^2}{E_j}.$$

Let d denote the number of components of θ. If the observed value of the chi-square statistic in step 5 is larger than the $(1-\alpha)$th quantile of a chi-square distribution with degrees of freedom $m - d - 1$, then we reject the hypothesis that the sample is from the discrete distribution with pmf $f(k;\, \theta)$ at the level of significance α.

If we have a large sample from a continuous distribution, then the chi-square goodness-of-fit test can be used to test the hypothesis that the sample is from a particular continuous distribution $F(x|\theta)$. The interval (the smallest observation, the largest observation) is divided into l subintervals, and the number O_j of data values fall in the jth interval is counted for $j = 1, \ldots, l$. The theoretical probability p_j that the underlying random variable assumes a value in the jth interval can be estimated using the distribution function $F(x|\widehat{\theta})$. The expected frequency for the jth interval can be computed as $E_j = p_j \times n$, for $j = 1, \ldots, l$. The chi-square statistic can be computed as in Step 5, and compared with the $(1-\alpha)$th quantile of the chi-square distribution with degrees of freedom $l - d - 1$, where d is the number of components of θ. If the computed value of the chi-square statistic is greater than the percentile, then the hypothesis will be rejected at the level of significance α.

1.5 Methods of Estimation

We shall describe here two classical methods of estimation, namely, the moment estimation and the method of maximum likelihood estimation. Let $X_1, \ldots, X_n$ be a sample of observations from a population with the distribution function $F(x|\theta_1, \ldots, \theta_k)$, where $\theta_1, \ldots, \theta_k$ are unknown parameters to be estimated based on the sample.

1.5.1 Moment Estimation

Let $f(x|\theta_1, \ldots, \theta_k)$ denote the pdf or pmf of a random variable X with cdf $F(x|\theta_1, \ldots, \theta_k)$. The moments about the origin are usually functions of $\theta_1, \ldots, \theta_k$. Notice that $E(X_i^k) = E(X_1^k)$, $i = 2, \ldots, n$, because the X_i's are identically distributed. The moment estimators can be obtained by solving the following system of equations for $\theta_1, \ldots, \theta_k$:

$$\begin{aligned}
\frac{1}{n}\sum_{i=1}^{n} X_i &= E(X_1)\\
\frac{1}{n}\sum_{i=1}^{n} X_i^2 &= E(X_1^2)\\
&\vdots\\
\frac{1}{n}\sum_{i=1}^{n} X_i^k &= E(X_1^k),
\end{aligned}$$

where

$$E(X_1^j) = \int_{-\infty}^{\infty} x^j f(x|\theta_1, \ldots, \theta_k)dx, \quad j = 1, 2, \ldots, k.$$

1.5.2 Maximum Likelihood Estimation

For a given sample $\mathbf{x} = (x_1, \ldots, x_n)$, the function defined by

$$L(\theta_1, \ldots, \theta_k |\, x_1, \ldots, x_n) = \prod_{i=1}^{n} f(x_i|\theta_1, \ldots, \theta_k)$$

is called the *likelihood function*. The maximum likelihood estimators are the values of $\theta_1, \ldots, \theta_k$ that maximize the likelihood function.

1.6 Inference

Let $\mathbf{X} = (X_1, \ldots, X_n)$ be a random sample from a population, and let $\mathbf{x} = (x_1, \ldots, x_n)$, where x_i is an observed value of X_i, $i=$ 1,...,n. For simplicity, let us assume that the distribution function $F(x|\theta)$ of the population depends only on a single parameter θ. In the sequel, $P(X \le x|\theta)$ means the probability that X is less than or equal to x when θ is the parameter of the distribution of X.

1.6.1 Hypothesis Testing

Some Terminologies

The main purpose of the hypothesis testing is to identify the range of the values of the population parameter based on a sample data. Let Θ denote the parameter space. The usual format of the hypotheses is

$$H_0 : \theta \in \Theta_0 \quad \text{vs.} \quad H_a : \theta \in \Theta_0^c, \tag{1.6.1}$$

where H_0 is called the null hypothesis, H_a is called the alternative or research hypothesis, Θ_0^c denotes the complement set of Θ_0, and $\Theta_0 \cup \Theta_0^c = \Theta$. For example, we want to test θ – the mean difference between durations of two treatments for a specific disease. If it is desired to compare these two treatment procedures, then one can set hypotheses as $H_0 : \theta = 0$ vs. $H_a : \theta \neq 0$.

In a hypothesis testing, decision based on a sample of data is made as to "reject H_0 and decide H_a is true" or "do not reject H_0." The subset of the sample space for which H_0 is rejected is called the *rejection region* or *critical region*. The complement of the rejection region is called the *acceptance region*.

Test Statistic: A statistic that is used to develop a test for the parameter of interest is called the test statistic. For example, usually the sample mean $\bar{X}$ is used to test about the mean of a population, and the sample proportion is used to test about the proportion in a population.

Errors and Powers

Type I Error: Wrongly rejecting H_0 when it is actually true is called the Type I error. Probability of making a Type I error while testing hypotheses is given by

$$P(\mathbf{X} \in R | \theta \in \Theta_0),$$

where R is the rejection region.

Type II Error: Wrongly accepting H_0 when it is false is called the Type II error. Probability of making a Type II error is given by

$$P(\mathbf{X} \in R^c | \theta \in \Theta_0^c),$$

where R^c denotes the acceptance region of the test.

Level of Significance: The maximum probability of making Type I error is called the level or level of significance; this is usually specified (common choices are 0.1, 0.05 or 0.01) before carrying out a test.

Power function: The power function $\beta(\theta)$ is defined as the probability of rejecting null hypothesis. That is,

$$\beta(\theta) = P(\mathbf{X} \in R | \theta \in \Theta)$$

Power: Probability of not making Type II error is called the power. That is, the probability of rejecting false H_0, and it can be expressed as $\beta(\theta) = P(\mathbf{X} \in R | \theta \in \Theta_0^c)$.

Size of a Test: The probability of rejecting H_0 at a given $\theta_1 \in \Theta_0$ is called the size at θ_1. That is, $P(\mathbf{X} \in R | \theta_1 \in \Theta_0)$ is called the size.

Level α Test: For a test, if $\sup_{\theta \in \Theta_0} P(X \in R | \theta) \leq \alpha$, then the test is called a level α test. That is, if the maximum probability of rejecting a true null hypothesis is less than or equal to α, then the test is called a level α test.

If the size exceeds α for some $\theta \in \Theta_0$, then the test is referred to as a *liberal* or *anti-conservative* test. If the sizes of the test are smaller than α, then it is referred to as a *conservative* test.

Size α Test: For a test, if $\sup_{\theta \in \Theta_0} P(X \in R | \theta) = \alpha$, then the test is called a size α test.

Unbiased Test: A test is said to be unbiased if $\beta(\theta_1) \leq \beta(\theta_2)$ for every θ_1 in Θ_0 and θ_2 in Θ_0^c.

A popular method of developing a test procedure is described below.

The Likelihood Ratio Test (LRT): Let $\mathbf{X} = (X_1, ..., X_n)$ be a random sample from a population with the pdf $f(x|\theta)$. Let $\mathbf{x} = (x_1, ..., x_n)$ be an observed sample. Then the likelihood function is given by

$$L(\theta | \mathbf{x}) = \prod_{i=1}^{n} f(x_i | \theta).$$

The LRT statistic for testing (1.6.1) is given by

$$\lambda(\mathbf{x}) = \frac{\sup_{\Theta_0} L(\theta|\mathbf{x})}{\sup_{\Theta} L(\theta|\mathbf{x})}.$$

Notice that $0 < \lambda(\mathbf{x}) < 1$, and the LRT rejects H_0 in (1.6.1) for smaller values of $\lambda(\mathbf{x})$.

Inferential procedures are usually developed based on a statistic $T(\mathbf{X})$ called pivotal quantity. The distribution of $T(\mathbf{X})$ can be used to make inferences on θ. The distribution of $T(\mathbf{X})$ when $\theta \in \Theta_0$ is called the *null distribution*, and when $\theta \in \Theta^c$ it is called the *non-null distribution*. The value $T(\mathbf{x})$ is called the observed value of $T(\mathbf{X})$. That is, $T(\mathbf{x})$ is the numerical value of $T(\mathbf{X})$ based on the observed sample $\mathbf{x}$.

P–Value: The p-value of a test is a measure of sample evidence in support of H_a. The smaller the p–value, the stronger the evidence for rejecting H_0. The p–value based on a given sample $\mathbf{x}$ is a constant in (0,1) whereas the p–value based on a random sample $\mathbf{X}$ is a uniform(0, 1) random variable. A level α test rejects H_0 whenever the p–value is less than or equal to α.

We shall now describe a test about θ based on a pivotal quantity $T(\mathbf{X})$. Consider testing the hypotheses

$$H_0 : \theta \le \theta_0 \quad \text{vs.} \quad H_a : \theta > \theta_0, \tag{1.6.2}$$

where θ_0 is a specified value. Suppose the statistic $T(\mathbf{X})$ is a stochastically increasing function of θ. That is, $T(\mathbf{X})$ is more likely to be large for large values of θ. The p–value for the hypotheses in (1.6.2) is given by

$$\sup_{\theta \le \theta_0} P\left(T(\mathbf{X}) > T(\mathbf{x})|\theta\right) = P\left(T(\mathbf{X}) > T(\mathbf{x})|\theta_0\right).$$

For two-sided alternative hypothesis, that is,

$$H_0 : \theta = \theta_0 \quad \text{vs.} \quad H_a : \theta \ne \theta_0,$$

the p–value is given by

$$2 \min\left\{P\left(T(\mathbf{X}) > T(\mathbf{x})|\theta_0\right), P\left(T(\mathbf{X}) < T(\mathbf{x})|\theta_0\right)\right\}.$$

For testing (1.6.2), let the critical point c be determined so that

$$\sup_{\theta \in \Theta_0} P(T(\mathbf{X}) \ge c|\theta) = \alpha.$$

Notice that H_0 will be rejected whenever $T(\mathbf{x}) > c$, and the region $\{\mathbf{x} : T(\mathbf{x}) > c\}$ is the rejection region.

The power function of the test for testing (1.6.2) is given by

$$\beta(\theta) = P(T(\mathbf{X}) > c|\theta).$$

The value $\beta(\theta_1)$ is the power at θ_1 if $\theta_1 \in \Theta_0^c$, and the value of $\beta(\theta_1)$ when $\theta_1 \in \Theta_0$ is the size at θ_1.

For an efficient test, the power function should be an increasing function of $|\theta - \theta_0|$ and/or the sample size. Between two level α tests, the one that has more power than the other should be used for practical applications.

1.6.2 Interval Estimation

Confidence Intervals

Let $L(\mathbf{X})$ and $U(\mathbf{X})$ be functions satisfying $L(\mathbf{X}) < U(\mathbf{X})$ for all samples. Consider the interval $(L(\mathbf{X}), U(\mathbf{X}))$. The probability

$$P((L(\mathbf{X}), U(\mathbf{X})) \text{ contains } \theta|\theta)$$

is called the *coverage probability* of the interval $(L(\mathbf{X}), U(\mathbf{X}))$. The minimum coverage probability, that is,

$$\inf_{\theta\in\Theta} P((L(\mathbf{X}), U(\mathbf{X})) \text{ contains } \theta|\theta)$$

is called the *confidence coefficient.* If the confidence coefficient is specified as, say, $1-\alpha$, then the interval $(L(\mathbf{X}), U(\mathbf{X}))$ is called a $1-\alpha$ confidence interval. That is, an interval is said to be a $1-\alpha$ confidence interval if its minimum coverage probability is $1-\alpha$.

One-Sided Limits: If the confidence coefficient of the interval $(L(\mathbf{X}), \infty)$ is $1-\alpha$, then $L(\mathbf{X})$ is called a $1-\alpha$ lower limit for θ, and if the confidence coefficient of the interval $(-\infty, U(\mathbf{X}))$ is $1-\alpha$, then $U(\mathbf{X})$ is called a $1-\alpha$ upper limit for θ.

Prediction Intervals

Prediction interval, based on a sample from a population with distribution $F(x|\theta)$, is constructed to assess the characteristic of an individual in the population. Let $\mathbf{X} = (X_1, ..., X_n)$ be a sample from $F(x|\theta)$. A $1-\alpha$ prediction interval for $X \sim F(x|\theta)$, where X is independent of $\mathbf{X}$, is a random interval $(L(\mathbf{X}), U(\mathbf{X}))$ that satisfies

$$\inf_{\theta\in\Theta} P((L(\mathbf{X}), U(\mathbf{X})) \text{ contains } X|\theta) = 1-\alpha.$$

The prediction interval for a random variable X is wider than the confidence interval for θ because it involves the uncertainty in estimates of θ and the uncertainty in X.

Tolerance Intervals

A p content – $(1-\alpha)$ coverage tolerance interval is an interval that would contain at least proportion p of the population measurements with confidence

$1-\alpha$. Let $\mathbf{X} = (X_1, ..., X_n)$ be a sample from $F(x|\theta)$, and $X \sim F(x|\theta)$ independently of $\mathbf{X}$. Then, a p content – $(1-\alpha)$ coverage tolerance interval is an interval $(L(\mathbf{X}), U(\mathbf{X}))$ that satisfies

$$P_{\mathbf{X}} \{P_X [L(\mathbf{X}) \le X \le U(\mathbf{X})] \ge p|\mathbf{X}\} = 1-\alpha.$$

One-sided tolerance limits are constructed similarly. For example, a statistic $L(\mathbf{X})$ is called a p content – $(1-\alpha)$ coverage lower tolerance limit, if it satisfies

$$P_{\mathbf{X}} \{P_X [X \ge L(\mathbf{X})] \ge p|\mathbf{X}\} = 1-\alpha.$$

1.7 Random Number Generation

The Inverse Method

The basic method of generating random numbers from a distribution is known as the inverse method. The inverse method for generating random numbers from a continuous distribution $F(x|\theta)$ is based on *the probability integral transformation*: If a random variable X follows $F(x|\theta)$, then the random variable $U = F(X|\theta)$ follows a uniform(0, 1) distribution. Therefore, if $U_1, \ldots, U_n$ are random numbers generated from uniform(0, 1) distribution, then

$$X_1 = F^{-1}(U_1, \theta), \ldots, X_n = F^{-1}(U_n, \theta)$$

are random numbers from the distribution $F(x|\theta)$. Thus, the inverse method is quite convenient if the inverse distribution function is easy to compute. For example, the inverse method is simple to use for generating random numbers from the Cauchy, Laplace, Logistic, and Weibull distributions.

If X is a discrete random variable with support $x_1 < x_2 < \ldots < x_n$ and cdf $F(x)$, then random variates can be generated as follows:

Generate a $U \sim$ uniform(0,1)
If $F(x_i) < U \le F(x_{i+1})$, set $X = x_{i+1}$.

X is a random number from the cdf $F(x)$. The above method should be used with the convention that $F(x_0) = 0$.

The Accept/Reject Method

Suppose that X is a random variable with pdf $f(x)$ and Y is a random variable with pdf $g(y)$. Assume that X and Y have common support, and random numbers from $g(y)$ can be easily generated. Define

$$M = \sup_x \frac{f(x)}{g(x)}.$$

The random numbers from $f(x)$ can be generated as follows.

1 Generate $U \sim$ uniform(0,1), and Y from $g(y)$
If $U < \frac{f(Y)}{Mg(Y)}$, deliver $X = Y$
else goto 1.

The expected number of trials required to generate one X is M.

1.8 Some Special Functions

In this section, some special functions which are used in the following chapters are given.

Gamma Function: The gamma function is defined by

$$\Gamma(x) = \int_0^\infty e^{-t} t^{x-1} dt \quad \text{for} \quad x > 0.$$

The gamma function satisfies the relation that $\Gamma(x+1) = x\Gamma(x)$.

Digamma Function: The digamma function is defined by

$$\psi(z) = \frac{d\left[\ln \Gamma(z)\right]}{dz} = \frac{\Gamma'(z)}{\Gamma(z)},$$

where

$$\Gamma(z) = \int_0^\infty e^{-t} t^{z-1} dt.$$

The value of $\gamma = -\psi(1)$ is called Euler's constant and is given by

$$\gamma = 0.5772 \;\; 1566 \;\; 4901 \;\; 5328 \;\; 6060 \cdots$$

For an integer $n \geq 2$, $\psi(n) = -\gamma + \sum\limits_{k=1}^{n-1} \frac{1}{k}$. Furthermore, $\psi(0.5) = -\gamma - 2\ln(2)$ and

$$\psi(n + 1/2) = \psi(0.5) + 2\left(1 + \frac{1}{3} + \cdots + \frac{1}{2n-1}\right), \quad n \geq 1.$$

The digamma function is also called the *Psi* function.

Beta Function: For $a > 0$ and $b > 0$, the beta function is defined by

$$B(a,b) = \frac{\Gamma(a)\Gamma(b)}{\Gamma(a+b)}.$$

The following logarithmic gamma function can be used to evaluate the beta function.

Logarithmic Gamma Function: The function $\ln\Gamma(x)$ is called the logarithmic gamma function, and it has wide applications in statistical computation. In particular, as shown in the later chapters, $\ln\Gamma(x)$ is needed in computing many distribution functions and inverse distribution functions. The following continued fraction for $\ln\Gamma(x)$ is quite accurate for $x \geq 8$ (see Hart et. al. 1968). Let

$b_0 = 8.33333333333333E-2$, $b_1 = 3.33333333333333E-2$,
$b_2 = 2.52380952380952E-1$, $b_3 = 5.25606469002695E-1$,
$b_4 = 1.01152306812684$, $b_5 = 1.51747364915329$,
$b_6 = 2.26948897420496$ and $b_7 = 3.00991738325940$.

Then, for $x \geq 8$,

$$\begin{aligned}\ln\Gamma(x) &= (x-0.5)\ln(x) - x + 9.1893853320467E-1 \\ &+ b_0/(x+b_1/(x+b_2/(x+b_3/(x+b_4/(x+b_5/(x+b_6/(x+b_7))))))).\end{aligned}$$

Using the above expression and the relation that $\Gamma(x+1) = x\Gamma(x)$, $\ln\Gamma(x)$ can be evaluated for $x < 8$ as

$$\begin{aligned}\ln\Gamma(x) &= \ln\Gamma(x+8) - \ln\prod_{i=0}^{7}(x+i) \\ &= \ln\Gamma(x+8) - \sum_{i=0}^{7}\ln(x+i).\end{aligned}$$

The following Fortran function subroutine based on the above method evaluates $\ln\Gamma(x)$ for a given $x > 0$.

```
      double precision function alng(x)
      implicit double precision (a-h, o-z)
      double precision b(8)
      logical check
      data b/8.33333333333333d-2, 3.33333333333333d-2,
     &       2.52380952380952d-1, 5.25606469002695d-1,
```

```
&         1.01152306812684d0,  1.51747364915329d0,
&         2.26948897420496d0,  3.00991738325940d0/
   if(x .lt. 8.0d0) then
      xx = x + 8.0d0
      check = .true.
   else
      check = .false.
      xx = x
   end if

   fterm = (xx-0.5d0)*dlog(xx) - xx + 9.1893853320467d-1
   sum = b(1)/(xx+b(2)/(xx+b(3)/(xx+b(4)/(xx+b(5)/(xx+b(6)
&        /(xx+b(7)/(xx+b(8))))))))
   alng = sum + fterm
   if(check)  alng = alng-dlog(x+7.0d0)-dlog(x+6.0d0)-dlog
&        (x+5.0d0)-dlog(x+4.0d0)-dlog(x+3.0d0)-dlog(x+2.0d0)
&        -dlog(x+1.0d0)-dlog(x)
   end
```

Chapter 2

Discrete Uniform Distribution

2.1 Description

The probability mass function of a discrete uniform random variable X is given by

$$P(X = k) = \frac{1}{N}, \quad k = 1, \ldots, N.$$

The cumulative distribution function is given by

$$P(X \leq k) = \frac{k}{N}, \quad k = 1, \ldots, N.$$

This distribution is used to model experimental outcomes which are "equally likely." The mean and variance can be obtained using the formulas that

$$\sum_{i=1}^{k} i = \frac{k(k+1)}{2} \quad \text{and} \quad \sum_{i=1}^{k} i^2 = \frac{k(k+1)(2k+1)}{6}.$$

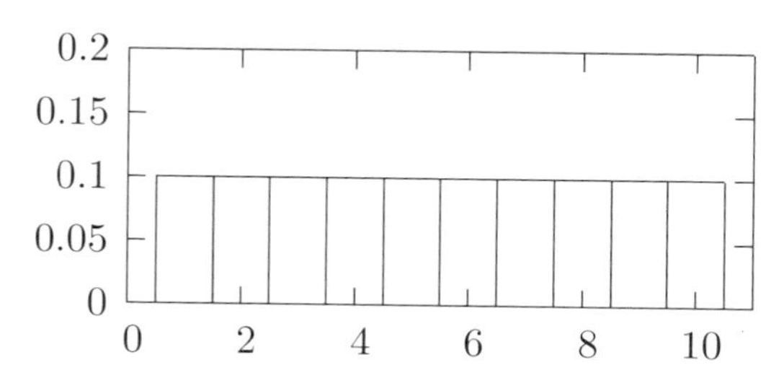

Figure 2.1 The Probability Mass Function when $N = 10$

2.2 Moments

Mean:	$\frac{N+1}{2}$
Variance:	$\frac{(N-1)(N+1)}{12}$
Coefficient of Variation:	$\left(\frac{N-1}{3(N+1)}\right)^{\frac{1}{2}}$
Coefficient of Skewness:	0
Coefficient of Kurtosis:	$3 - \frac{6(N^2+1)}{5(N-1)(N+1)}$
Moment Generating Function:	$M_X(t) = \frac{e^t(1-e^{Nt})}{N(1-e^t)}$
Mean Deviation:	$\begin{cases} \frac{N^2-1}{4N} & \text{if } N \text{ is odd,} \\ \frac{N}{4} & \text{if } N \text{ is even.} \end{cases}$

Chapter 3

Binomial Distribution

3.1 Description

A binomial experiment involves n independent and identical trials such that each trial can result in to one of the two possible outcomes, namely, success or failure. If p is the probability of observing a success in each trial, then the number of successes X that can be observed out of these n trials is referred to as the binomial random variable with n trials and success probability p. The probability of observing k successes out of these n trials is given by the probability mass function

$$P(X = k|n, p) = \binom{n}{k} p^k (1-p)^{n-k}, \quad k = 0, 1, ..., n. \tag{3.1.1}$$

The cumulative distribution function of X is given by

$$P(X \le k|n, p) = \sum_{i=0}^{k} \binom{n}{i} p^i (1-p)^{n-i}, \quad k = 0, 1, ..., n. \tag{3.1.2}$$

Binomial distribution is often used to estimate or determine the proportion of individuals with a particular attribute in a large population. Suppose that a random sample of n units is drawn by sampling with replacement from a finite population or by sampling without replacement from a large population. The number of units that contain the attribute of interest in the sample follows a binomial distribution. The binomial distribution is not appropriate if the sample was drawn without replacement from a small finite population; in this situation the hypergeometric distribution in Chapter 4 should be used. For practical purposes, binomial distribution can be used for a population of size around 5,000 or more.

We denote a binomial distribution with n trials and success probability p by binomial(n, p). This distribution is right-skewed when $p < 0.5$, and left-skewed when $p > 0.5$ and symmetric when $p = 0.5$. See the plots of probability mass functions in Figure 3.1. For large n, binomial distribution is approximately symmetric about its mean np.

3.2 Moments

Mean:	np
Variance:	$np(1-p)$
Mode:	The largest integer $\leq (n+1)p$
Mean Deviation:	$2n\binom{n-1}{m}p^{m+1}(1-p)^{n-m}$, where m denotes the largest integer $\leq np$. [Kamat 1965]
Coefficient of Variation:	$\sqrt{\frac{1-p}{np}}$
Coefficient of Skewness:	$\frac{1-2p}{\sqrt{np(1-p)}}$
Coefficient of Kurtosis:	$3 - \frac{6}{n} + \frac{1}{np(1-p)}$
Factorial Moments:	$E\left(\prod_{i=1}^{k}(X-i+1)\right) = p^k \prod_{i=1}^{k}(n-i+1)$
Moments about the Mean:	$np(1-p)\sum_{i=0}^{k-2}\binom{k-1}{i}\mu_i - p\sum_{i=0}^{k-2}\binom{k-1}{i}\mu_{i+1}$, where $\mu_0 = 1$ and $\mu_1 = 0$. [Kendall and Stuart 1958, p. 122]
Moments Generating Function:	$(pe^t + (1-p))^n$
Probability Generating Function:	$(pt + (1-p))^n$

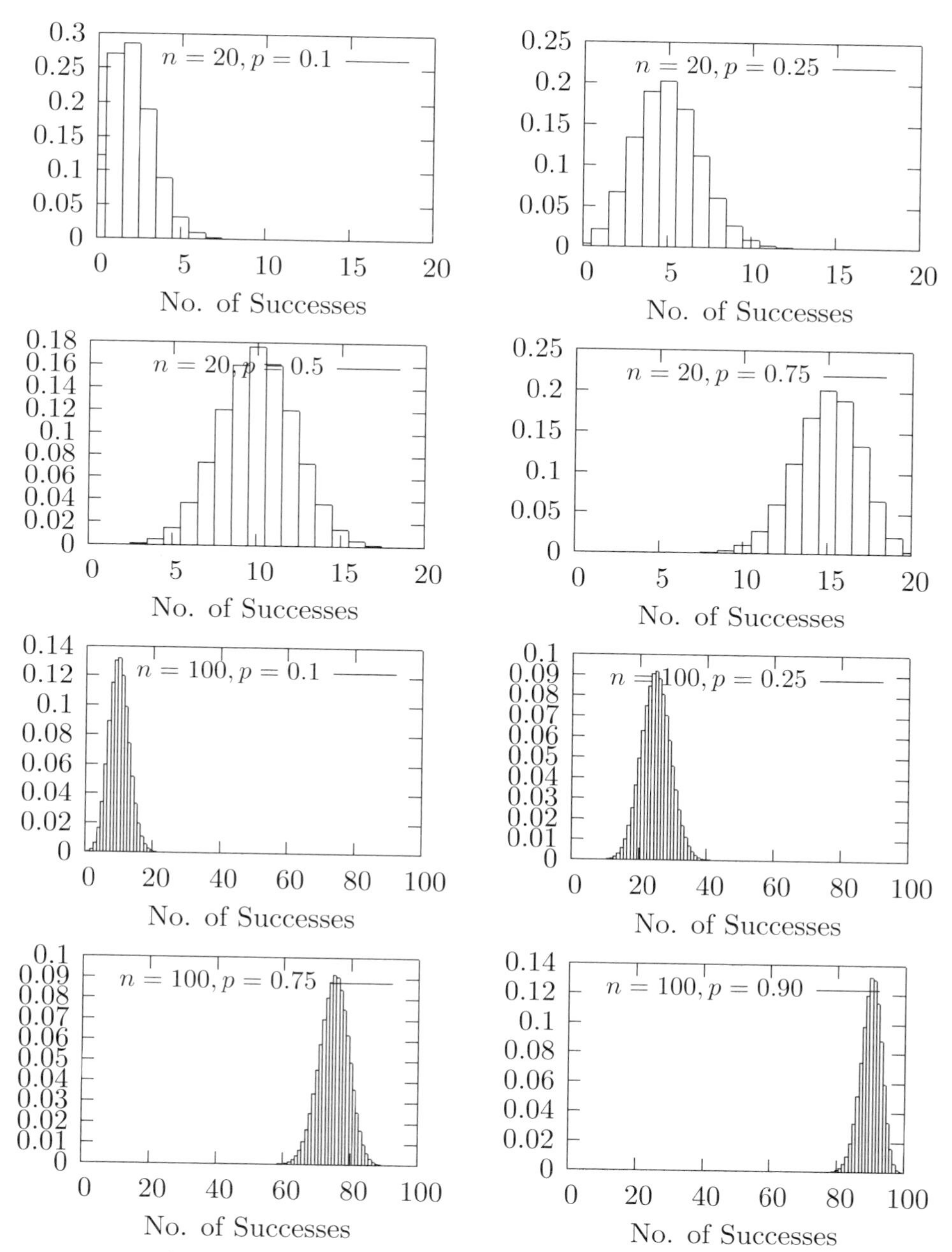

Figure 3.1 Binomial Probability Mass Functions

3.3 Computing Table Values

The dialog box [StatCalc→Discrete→Binomial] computes the following table values.

1. Tail Probabilities, Critical Values, and Moments.
2. Test for Proportion and Power Calculation [Section 3.4].
3. Confidence Interval for Proportion and Sample Size for Precision [Section 3.5].
4. Test for the Difference between Two Proportions and Power Calculation [Section 3.6].
5. P-values for Fisher's Exact Test and Power Calculation [Section 3.7].

The dialog [StatCalc→Discrete→Binomial→Probabilities, Critical Values and Moments] can be used to compute the following.

To compute probabilities: Enter the values of the number of trials n, success probability p, and the observed number of successes k; click [P].

Example 3.3.1 When $n = 20, p = 0.2$, and $k = 4$,

$$P(X \leq 4) = 0.629648,\ P(X \geq 4) = 0.588551 \text{ and } P(X = 4) = 0.218199.$$

To compute the value of p: Input values for the number of trials n, the number of successes k and for the cumulative probability P(X <= k); click [s].

Example 3.3.2 When $n = 20$, $k = 4$ and $P(X \leq k) = 0.7$, the value of p is 0.183621.

To compute the value of n: Enter the values of p, the number of successes k, and P(X <= k); click [n].

Example 3.3.3 When $p = 0.20$, $k = 6$ and $P(X \leq k) = 0.4$, the value of n is 36.

To compute the value of k: Enter the values of n, p, and the cumulative probability P(X <= k); click [k]. If the cumulative probability c is greater than 0.5, then *StatCalc* computes the smallest value of k such that $P(X \geq k) \leq 1 - c$. That is, the value of k is computed so that the right-tail probability is less than or equal to $1 - c$; if $c < 0.5$, then the largest value of k is computed so that the $P(X \leq k) \leq c$.

Example 3.3.4 When $p = 0.4234$, $n = 43$ and $P(X <= k) = 0.90$, the value of k is 23. Notice that $P(X \geq 23) = 0.0931953$, which is less than $1 - 0.90 = 0.10$.

If $P(X \leq k) = 0.10$, then k is 13, and $P(X \leq 13) = 0.071458$. Note that $P(X \leq 14) = 0.125668$ which is greater than 0.10.

To compute moments: Enter values for n and p; click [M].

Illustrative Examples

Example 3.3.5 Suppose that a balanced coin is to be flipped 20 times. Find the probability of observing

a. ten heads;

b. at least 10 heads;

c. between 8 and 12 heads.

[StatCalc→Discrete→Binomial→Probabilities, Critical Values and Moments]
Solution: Let X denote the number of heads that can be observed out of these 20 flips. Here, the random variable X is binomial with $n = 20$, and the success probability $= 0.5$, which is the probability of observing a head at each flip.

a. To find the probability, enter 20 for n, 0.5 for success probability, 10 for k, and click on [P] to get $P(X = 10) = 0.176197$. That is, the chances of observing exactly 10 heads are about 18%.

b. To get this probability, enter 20 for n, 0.5 for p, 10 for k, and click [P] to get $P(X \geq 10) = 0.588099$. That is, the chances of observing 10 or more heads are about 59%.

c. The desired probability is

$$\begin{aligned} P(8 \leq X \leq 12) &= P(X \leq 12) - P(X \leq 7) \\ &= 0.868412 - 0.131588 \\ &= 0.736824. \end{aligned}$$

Example 3.3.6 What are the chances of observing exactly 3 girls in a family of 6 children?

[StatCalc→Discrete→Binomial→Probabilities, Critical Values and Moments]
Solution: Let us assume that the probability of giving birth to a boy = probability of giving birth to a girl = 0.5. Let X be the number of girls in the family. Here, X is a binomial random variable with $n = 6$ and $p = 0.5$. To find the probability, enter 6 for n, 0.5 for p, and 3 for k; click [P] to get $P(X = 3) = 0.3125$.

3.4 Test for the Proportion

Suppose that investigation of a sample of n units from a population revealed that k units have a particular attribute. Let p denote the proportion of individuals in the population with the attribute. The following inferential procedures for p are based on n and k.

3.4.1 An Exact Test

For testing

$$H_0 : p \le p_0 \quad \text{vs.} \quad H_a : p > p_0, \tag{3.4.1}$$

the null hypothesis will be rejected if the p-value $P(X \ge k|n, p_0) \le \alpha$, for testing

$$H_0 : p \ge p_0 \quad \text{vs.} \quad H_a : p < p_0, \tag{3.4.2}$$

the null hypothesis will be rejected if the p-value $P(X \le k|n, p_0) \le \alpha$, and for testing

$$H_0 : p = p_0 \quad \text{vs.} \quad H_a : p \ne p_0, \tag{3.4.3}$$

the null hypothesis will be rejected if the p-value

$$2\min\{P(X \le k|n, p_0), P(X \ge k|n, p_0)\} \le \alpha. \tag{3.4.4}$$

3.4.2 Power of the Exact Test

For a right-tail test (i.e., testing hypotheses in (3.4.1)), the exact power can be computed using the expression

$$\sum_{k=0}^{n} \binom{n}{k} p^k (1-p)^{n-k} I\left(P\left(X \ge k|n, p_0\right) \le \alpha\right), \tag{3.4.5}$$

where $I(.)$ is the indicator function. A power expression for a left-tail test (i.e., testing hypotheses in (3.4.2)) can be obtained by replacing $P(X \ge k|n, p_0)$ in (3.4.5) by $P(X \le k|n, p_0)$.

For a two-tail test (i.e., testing hypotheses in (3.4.3)), the exact power can be computed using the expression

$$\sum_{k=0}^{n} \binom{n}{k} p^k (1-p)^{n-k} I\left(2\min\{P(X \le k|n, p_0), P(X \ge k|n, p_0)\} \le \alpha\right). \tag{3.4.6}$$

The random variable X in (3.4.5) and (3.4.6) is a binomial(n, p_0) random variable and $p \neq p_0$. Letting $p = p_0$ in the above expressions, we can get the Type I error rate (size) of the test.

For a given n, k, and p_0, the dialog box [StatCalc→Discrete→Binomial→Test for p and Sample Size for Power] computes the p-values for testing binomial proportion; for a given p, p_0, α and power, this dialog box also computes the sample size required to attain a specified power.

Example 3.4.1 (*Calculation of p-values*) When $n = 20$, $k = 8$, and $p_0 = 0.2$, it is desired to test (3.4.1) at the level of 0.05. To compute the p-value, select [StatCalc→Discrete→Binomial→Test for p and Sample Size for Power], and enter the values of n, k and p_0 in the dialog box; click [p-values for] to get 0.0321427. If the nominal level is 0.05, then the null hypothesis H_0 in (3.4.1) will be rejected in favor of the alternative hypothesis H_a in (3.4.1). However, at the same nominal level, the H_0 in (3.4.3) can not be rejected because the p-value for this two-tail test is 0.0642853, which is not less than 0.05.

Example 3.4.2 (*Calculation of p-values*) A pharmaceutical company claims that 75% of doctors prescribe one of its drugs for a particular disease. In a random sample of 40 doctors, 23 prescribed the drug to their patients. Does this information provide sufficient evidence to indicate that the actual percentage of doctors who prescribe the drug is less than 0.75? Test at the level 0.05.

Solution: Let p be the actual proportion of doctors who prescribe the drug to their patients. We want to test

$$H_0 : p \geq 0.75 \quad \text{vs.} \quad H_a : p < 0.75.$$

To compute the p-value for testing above hypotheses, select the dialog box [StatCalc→Discrete→Binomial→Test for p and Sample Size for Power], enter 40 for n, 23 for observed k, and 0.75 for [Value of p0]. Click on [p-values for]. The p-value for the above left-tail test is 0.0115614, which is less than 0.05. Thus, we can conclude, on the contrary to the manufacturer's claim, that less than 75% of doctors prescribe the drug.

For a given n, population proportion p, and p_0, the dialog box [StatCalc→Discrete →Binomial→Test for p and Sample Size for Power] also computes the power of the test.

Example 3.4.3 When $n = 35$, $p_0 = 0.2$, nominal level $= 0.05$, and $p = 0.4$, the power of the test for the hypotheses in (3.4.1) is 0.804825. To compute the power, enter 1 to indicate right-tail test, 0.05 for the level, 0.2 for [Value of p0], 0.4 for [Guess p], and 35 for [S Size]; click on [Power]. For the hypotheses in

(3.4.3), the power is 0.69427; to compute this power, enter 3 to indicate two-tail test, and click [Power].

Example 3.4.4 (Sample Size Calculation) Suppose that a researcher believes that a new drug is 20% more effective than the existing drug, which has a success rate of 70%. The required sample size to test his belief (at the level 0.05 and power 0.90) can be computed as follows. Enter 1 to indicate right-tail test, 0.05 for the level, 0.7 for the value of p_0, 0.9 for [Guess p], 0.9 for [Power] and click on [S Size] to get 37; now click on [Power] to get 0.928915, which is the actual power when the sample size is 37.

3.5 Confidence Intervals for the Proportion

3.5.1 An Exact Confidence Interval

An exact confidence interval for a binomial proportion p can be obtained using the Clopper–Pearson (1934) approach. For a given sample size n and an observed number of successes k, the lower limit p_L for p is the solution of the equation

$$\sum_{i=k}^{n} \binom{n}{i} p_L^i (1 - p_L)^{n-i} = \frac{\alpha}{2},$$

and the upper limit p_U is the solution of the equation

$$\sum_{i=0}^{k} \binom{n}{i} p_U^i (1 - p_U)^{n-i} = \frac{\alpha}{2}.$$

Using a relation between the binomial and beta distributions (see Section 16.6.2), it can be shown that

$$p_L = \text{beta}^{-1}(\alpha/2; k, n-k+1) \quad \text{and} \quad p_U = \text{beta}^{-1}(1-\alpha/2; k+1, n-k),$$

where $\text{beta}^{-1}(c; a, b)$ denotes the cth quantile of a beta distribution with the shape parameters a and b. The interval (p_L, p_U) is an exact $1 - \alpha$ confidence interval for p, in the sense that the coverage probability is always greater than or equal the specified confidence level $1 - \alpha$.

One-sided $1 - \alpha$ lower limit for p is $\text{beta}^{-1}(\alpha; k, n-k+1)$ and one-sided $1 - \alpha$ upper limit for p is $\text{beta}^{-1}(1-\alpha; k+1, n-k)$. When $k = n$, the upper limit is 1 and the lower limit is $\alpha^{\frac{1}{n}}$; when $k = 0$, the lower limit is 0 and the upper limit is $1 - \alpha^{\frac{1}{n}}$.

Expected Length

For a given confidence level $1-\alpha$ and p, the expected length of (p_L, p_U) is given by

$$\sum_{k=0}^{n} \binom{n}{k} p^k (1-p)^{n-k} (p_U - p_L), \tag{3.5.1}$$

where p_L and p_U are as defined above.

The dialog box [StatCalc→Discrete→Binomial→CI for p and Sample Size for Precision] uses the above formulas to compute confidence intervals for p. This dialog box also compute the sample size required to estimate the population proportion within a given precision.

Remark 3.5.1 Suppose that a binomial experiment is repeated m times, and let k_j denote the number of successes, and let n_j denote the number of trials at the jth time, $j = 1, 2, ..., m$. Then, the above inferential procedures are valid with (n, k) replaced by $\left(\sum_{j=1}^{m} n_j, \sum_{j=1}^{m} k_j\right)$.

3.5.2 Computing Exact Limits and Sample Size Calculation

For a given n and k, the dialog box [StatCalc→Discrete→Binomial→CI for p and Sample Size for Precision] computes one-sided confidence limits and confidence intervals for p. Exact methods described in the preceding section are used for computation of the confidence intervals. Furthermore, this dialog box computes the required sample size for a given precision (that is, one half of the expected length in (3.5.1)).

Example 3.5.1 (*Confidence Intervals for p*) Suppose that a binomial experiment of 40 trials resulted in 5 successes. To find a 95% confidence interval, enter 40 for n, 5 for the observed number of successes k, and 0.95 for the confidence level; click [2-sided] to get (0.0419, 0.2680). For one-sided limits, click [1-sided] to get 0.0506 and 0.2450. That is, 95% one-sided lower limit for p is 0.0506, and 95% one-sided upper limit for p is 0.2450.

Example 3.5.2 (*Confidence Intervals for p*) The manufacturer of a product reports that only 5 percent or fewer of his products are defective. In a random sample of 25 such products, 4 of them were found to be defective. Find 95% confidence limits for the true percentage of defective products.

Solution: Let X be the number of defective products in the sample. Then, X is a binomial random variable with $n = 25$ and $p = 0.05$, which is the probability of observing a defective product in the lot. To get a 95% confidence interval for the actual percentage of defective products, select the dialog box [StatCalc→Discrete→Binomial→CI for p and Sample Size for Precision] from *StatCalc*, enter 25 for n, 4 for k, and 0.95 for the confidence level. Click on [2-sided] to get 0.0454 and 0.3608. This means that the actual percentage of defective items is somewhere between 4.5 and 36, with 95% confidence. Click [1-sided] to get the lower limit 0.05656; this means that the actual percentage of defective products is at least 5.66, with 95% confidence.

The dialog box [StatCalc→Discrete→Binomial→CI for p and Sample Size for Precision] also computes the sample size required to have an interval estimate for p within a given precision.

Example 3.5.3 (Sample Size Calculation) A researcher hypothesizes that the proportion of individuals with the attributes of interest in a population is 0.3, and he wants to estimate the true proportion within ±5% with 95% confidence. To compute the required sample size, enter 0.95 for [Conf Level], 0.3 for [Guess p] and 0.05 for [Half-Width]. Click on [Sam Size] to get 340.

3.6 A Test for the Difference between Two Proportions

Suppose that inspection of a sample of n_1 individuals from a population revealed k_1 units with a particular attribute, and a sample of n_2 individuals from another population revealed k_2 units with the same attribute. The problem of interest is to test the difference $p_1 - p_2$, where p_i, $i = 1, 2$, denotes the true proportion of individuals in the ith population with the attribute of interest.

3.6.1 An Unconditional Test

Let $\widehat{p}_i = k_i/n_i$, $i = 1, 2$. The p-value for testing hypotheses

$$H_0 : p_1 \le p_2 \quad \text{vs.} \quad H_a : p_1 > p_2$$

can be computed using the formula

$$P(k_1, k_2, n_1, n_2) = \sum_{x_1=0}^{n_1} \sum_{x_2=0}^{n_2} f(x_1|n_1, \widehat{p}_1) f(x_2|n_2, \widehat{p}_2)$$

$$\times \quad I\left(Z(x_1, x_2, n_1, n_2) \geq Z(k_1, k_2, n_1, n_2)\right), \quad (3.6.1)$$

where $I(.)$ is the indicator function,

$$f(x_i|n_i, \widehat{p}_i) = \binom{n_i}{x_i} \widehat{p}_x^{x_i - 1} (1 - \widehat{p}_x)^{n_i - x_i}, \; i = 1, 2,$$

$$Z(x_1, x_2, n_1, n_2) = \frac{x_1 - x_2}{\sqrt{\widehat{p}_x(1 - \widehat{p}_x)/(1/n_1 + 1/n_2)}}, \text{ and } \widehat{p}_x = \frac{x_1 + x_2}{n_1 + n_2}.$$

The terms $Z(k_1, k_2, n_1, n_2)$ and $\widehat{p}_k$ are respectively equal to $Z(x_1, x_2, n_1, n_2)$ and $\widehat{p}_x$ with x replaced by k. The null hypothesis will be rejected when the p-value in (3.6.1) is less than or equal to the nominal level α. This test is due to Storer and Kim (1990). Even though this test is approximate, its Type I error rates seldom exceed the nominal level and it is more powerful than Fisher's conditional test in Section 3.7.

The p-values for a left-tail test and two-tail test can be computed similarly.

3.6.2 Power of the Unconditional Test

For a given $(n_1, n_2, p_1, p_2, \alpha)$, the exact power of the unconditional test given above can be computed using the expression

$$\sum_{k_1=0}^{n_1} \sum_{k_2=0}^{n_2} f(k_1|n_1, p_1) f(k_2|n_2, p_2) I\left(P(k_1, k_2, n_1, n_2) \leq \alpha\right), \quad (3.6.2)$$

where the p-value $P(k_1, k_2, n_1, n_2)$ is given in (3.6.1). The powers of a left-tail test and a two-tail test can be computed similarly.

The dialog box [StatCalc→Discrete→Binomial→Test for p1 - p2 and Power Calculation] computes the p-value of the unconditional test described in the preceding paragraphs, and its exact power using (3.6.2).

Example 3.6.1 (P-values of the Unconditional Test) Suppose a sample of 25 observations from population 1 yielded 20 successes and a sample of 20 observations from population 2 yielded 10 successes. Let p_1 and p_2 denote the proportions of successes in populations 1 and 2, respectively. We want to test

$$H_0 : p_1 \leq p_2 \;\; \text{vs.} \;\; H_a : p_1 > p_2.$$

To compute the p-value, enter the numbers of successes and sample sizes, click on [p-values for] to get the p-value of 0.024635. The p-value for the two-tail test is 0.0400648.

The dialog box [StatCalc→Discrete→Binomial→Test for p1 - p2 and Power Calculation] also computes the power of the unconditional test for a given level of significance, guess values of population proportions p_1 and p_2, and sample sizes.

Example 3.6.2 (*Sample Size Calculation*) Suppose the sample size for each group needs to be determined to carry out a two-tail test at the level of significance $\alpha = 0.05$ and power $= 0.80$. Furthermore, the guess values of the proportions are given as $p_1 = 0.45$ and $p_2 = 0.15$. To determine the sample size, enter 2 for two-tail test, 0.05 for [Level], 0.45 for p_1, 0.15 for p_2, and 28 for each sample size. Click [Power] to get a power of 0.697916. Note that the sample size gives a power less than 0.80. This means that the sample size required to have a power of 0.80 is more than 28. Enter 32 (for example) for both sample sizes and click on [Power] radio button. Now the power is 0.752689. Again, raise the sample size to 36, and click the [Power] radio button. We now see that power is 0.81429, which is slightly higher than the desired power 0.80. Thus, the required sample size for each population to have a power of at least 0.80 is 36. Also, note that the power at $n_1 = n_2 = 35$ is 0.799666.

Remark 3.6.1 The power can also be computed for unequal sample sizes. For instance, when $n_1 = 30$, $n_2 = 42$, $p_1 = 0.45$, $p_2 = 0.15$, the power of a two-tail test at the nominal level 0.05 is 0.804295. For the same configuration, a power of 0.808549 can be attained if $n_1 = 41$ and $n_2 = 29$.

3.7 Fisher's Exact Test

Let X be a binomial(n_1, p_1) random variable, and Y be a binomial(n_2, p_2) random variable. Assume that X and Y are independent. When $p_1 = p_2$, the conditional probability of observing $X = k$, given that $X + Y = m$ is given by

$$P(X = k|X + Y = m) = \frac{\binom{n_1}{k}\binom{n_2}{m-k}}{\binom{n_1+n_2}{m}}, \quad \max\{0, m - n_2\} \leq k \leq \min\{n_1, m\}. \tag{3.7.1}$$

The pmf in (3.7.1) is known as the hypergeometric$(m, n_1, n_1 + n_2)$ pmf (see Section 4.1). This conditional distribution can be used to test the hypotheses regarding $p_1 - p_2$. For example, when

$$H_0 : p_1 \leq p_2 \ \text{ vs. } \ H_a : p_1 > p_2,$$

the null hypothesis will be rejected if the p-value $P(X \geq k|X + Y = m)$, which can be computed using (3.7.1), is less than or equal to the nominal level α. Similarly, the p-value for testing $H_0 : p_1 \geq p_2$ vs. $H_a : p_1 < p_2$ is given by $P(X \leq k|X + Y = m)$.

In the form of 2×2 table we have:

Sample	Successes	Failures	Totals
1	k	$n_1 - k$	n_1
2	$m - k$	$n_2 - m + k$	n_2
Totals	m	$n_1 + n_2 - m$	$n_1 + n_2$

3.7.1 Calculation of p-Values

For a given 2×2 table, the dialog box [StatCalc→Discrete→Binomial→Fisher's Exact Test and Power Calculation] computes the probability of observing k or more successes (as well as the probability of observing k or less number of successes) in the cell (1,1). If either probability is less than $\alpha/2$, then the null hypothesis of equal proportion will be rejected at the level α. Furthermore, for a given level α, sample sizes, and guess values on p_1 and p_2, this dialog box also computes the exact power. To compute the power, enter sample sizes, the level of significance, guess values on p_1 and p_2, and then click [Power].

Example 3.7.1 A physician believes that one of the causes of a particular disease is long-term exposure to a chemical. To test his belief, he examined a sample of adults and obtained the following 2×2 table:

Group	Symptoms Present	Symptoms Absent	Totals
Exposed	13	19	32
Unexposed	4	21	25
Totals	17	40	57

The hypotheses of interest are

$$H_0 : p_e \leq p_u \quad \text{vs.} \quad H_a : p_e > p_u,$$

where p_e and p_u denote, respectively, the actual proportions of exposed people and unexposed people who have the symptom. To find the p-value, select the dialog box [StatCalc→Discrete→Binomial→Fisher's Exact Test and Power Calculation], enter the cell frequencies and click [Prob <= (1,1) cell]. The p-value is 0.04056. Thus, at the 5% level, the data provide sufficient evidence to indicate that there is a positive association between the prolonged exposure to the chemical and the disease.

3.7.2 Exact Powers

The exact power of Fisher's test described in the preceding sections can be computed using the expression

$$\sum_{k_1=0}^{n_1}\sum_{k_2=0}^{n_2} f(k_1|n_1,p_1)f(k_2|n_2,p_2)I(\text{p}-\text{value}\le\alpha), \tag{3.7.2}$$

where $f(x|n,p)=\binom{n}{x}p^x(1-p)^{n-x}$, $I(.)$ is the indicator function and the p–value is given in (3.7.1). *StatCalc* uses the above formula for computing the power of Fisher's exact test.

Example 3.7.2 (Sample Size Calculation) Suppose that an experimenter wants to apply Fisher's exact test for testing two proportions. He believes that the proportion p_1 is about 0.3 more than the proportion $p_2 = 0.15$, and wants to compute the required sample sizes to have a power of 0.9 for testing

$$H_0: p_1 \le p_2 \text{ vs. } H_a: p_1 > p_2$$

at the level 0.05.

To determine the sample size required from each population, enter 28 (this is our initial guess) for both sample sizes, 0.45 $(= .3 + .15)$ for p_1, 0.15 for p_2, 0.05 for level, and click power to get 0.724359. This is less than 0.9. After trying a few values larger than 28 for each sample size, we find the required sample size is 45 from each population. In this case, the actual power is 0.90682.

Example 3.7.3 (Unconditional Test vs. Conditional Test) To understand the difference between the sample sizes needed for the unconditional test and Fisher's test, let us consider Example 3.6.2. Suppose the sample size for each group needs to be determined to carry out a two-tail test at the level of significance $\alpha = 0.05$ and power $= 0.80$. Furthermore, the guess values of the proportions are given as $p_1 = 0.45$ and $p_2 = 0.15$. To determine the sample size, enter 2 for two-tail test, 0.05 for [Level], 0.45 for p_1, 0.15 for p_2, and 28 for each sample size. Click [Power] to get a power of 0.612953. By raising the each sample size to 41, we get the power of 0.806208. Note that the power at $n_1 = n_2 = 40$ is 0.7926367.

If we decide to use the unconditional test given in Section 3.6, then the required sample size for each group is 36. Because the unconditional test is more powerful than Fisher's test, it requires smaller samples to attain the same power.

Remark 3.7.1 Note that the power can also be computed for unequal sample sizes. For instance, when $n_1 = 30$, $n_2 = 42$, $p_1 = 0.45$, $p_2 = 0.15$, the power of the two-tail test at a nominal level of 0.05 is 0.717820. For the same configuration, a power of 0.714054 can be attained if $n_1 = 41$ and $n_2 = 29$. For the same sample sizes, the unconditional test provides larger powers than those of Fisher's test (see Remark 3.6.1).

3.8 Properties and Results

3.8.1 Properties

1. Let $X_1, \ldots, X_m$ be independent random variables with $X_i \sim$ binomial(n_i, p), $i = 1, 2, ..., m$. Then

$$\sum_{i=1}^{m} X_i \sim \text{binomial}\left(\sum_{i=1}^{m} n_i, p\right).$$

2. Let X be a binomial(n, p) random variable. For fixed k,

$$P(X \le k|n, p)$$

is a nonincreasing function of p.

3. Recurrence Relations:

(i) $P(X = k+1) = \frac{(n-k)p}{(k+1)(1-p)} P(X = k), \quad k = 0, 1, 2, \ldots, n-1.$

(ii) $P(X = k-1) = \frac{k(1-p)}{(n-k+1)p} P(X = k), \quad k = 1, 2, \ldots, n.$

4. (i) $P(X \ge k) = p^k \sum_{i=k}^{n} \binom{i-1}{k-1} (1-p)^{i-k}.$

(ii) $\sum_{i=k}^{n} i\binom{n}{i} p^i (1-p)^{n-i} = npP(X \ge k) + k(1-p)P(X = k).$

[Patel et al. (1976), p. 201]

3.8.2 Relation to Other Distributions

1. Bernoulli: Let $X_1, \ldots, X_n$ be independent Bernoulli(p) random variables with success probability p. That is, $P(X_i = 1) = p$ and $P(X_i = 0) = 1 - p$, $i = 1, \ldots, n$. Then

$$\sum_{i=1}^{n} X_i \sim \text{binomial}(n, p).$$

2. Negative Binomial: See Section 7.7.2.

3. Hypergeometric: See Section 4.8.2.
4. F Distribution: See Section 12.4.2.
5. Beta: See Section 16.6.2.

3.8.3 Approximations

1. Let n be such that $np > 5$ and $n(1-p) > 5$. Then,

$$P(X \leq k|n,p) \simeq P\left(Z \leq \frac{k - np + 0.5}{\sqrt{np(1-p)}}\right),$$

and

$$P(X \geq k|n,p) \simeq P\left(Z \geq \frac{k - np - 0.5}{\sqrt{np(1-p)}}\right),$$

where Z is the standard normal random variable.

2. Let $\lambda = np$. Then, for large n and small p,

$$P(X \leq k|n,p) \simeq P(Y \leq k) = \sum_{i=0}^{k} \frac{e^{-\lambda}\lambda^i}{i!},$$

where Y is a Poisson random variable with mean λ.

3.9 Random Number Generation

```
Input:
      n = number of trials
      p = success probability
      ns = desired number of binomial random numbers
Output:
      x(1),...,x(ns) are random numbers from the binomial(n, p)
      distribution
```

Algorithm 3.9.1

The following algorithm, which generates the binomial(n, p) random number as the sum of n Bernoulli(p) random numbers, is satisfactory and efficient for small n.

```
    Set   k = 0
    For j = 1 to ns
       For i = 1 to n
         Generate u from uniform(0, 1)
         If u <= p,  k = k + 1
       [end i loop]
2      x(j)  = k
       k = 0
    [end j loop]
```

The following algorithm first computes the probability and the cumulative probability around the mode of the binomial distribution, and then searching for k sequentially so that $P(X \leq k-1) < u \leq P(X \leq k)$, where u is a uniform random variate. Depending on the value of the uniform variate, forward or backward search from the mode will be carried out. If n is too large, search for k may be restricted in the interval $np \pm c\sqrt{np(1-p)}$, where $c \geq 7$. Even though this algorithm requires the computation of the cumulative probability around the mode, it is accurate and stable.

Algorithm 3.9.2

```
Set k =  int((n + 1)*p)
    s = p/(1 - p)
    pk = P(X = k)
    df = P(X <= k)
    rpk = pk; rk = k;
For j = 1 to ns
    Generate u from uniform(0, 1)
    If u > df, go to 2
1   u = u + pk
    If k = 0 or u > df, go to 3
    pk = pk*k/(s*(n - k + 1))
    k = k - 1
    go to 1
2   pk = (n - k)*s*pk/(k + 1)
    u = u - pk
    k = k + 1
    If k = n or u <= df, go to 3
    go to 2
3   x(j) = k
    k = rk
```

```
    pk = rpk
[end j loop]
```

For other algorithms, see Kachitvichyanukul and Schmeiser (1988).

3.10 Computation of Probabilities

For small n, the probabilities can be computed in a straightforward manner. For large values of n, logarithmic gamma function $\ln\Gamma(x)$ (see Section 1.8) can be used to compute the probabilities.

To Compute $P(X = k)$:

Set $x = \ln\Gamma(n+1) - \ln\Gamma(k+1) - \ln\Gamma(n-k+1)$
$y = k * \ln(p) + (n-k) * \ln(1-p)$
$P(X = k) = \exp(x + y)$.

To Compute $P(X \le k)$:

Compute $P(X = k)$
Set $m = \mathrm{int}(np)$
If $k \le m$, compute $P(X = k - 1)$ using the backward recursion relation

$$P(X = k-1) = \frac{k(1-p)}{(n-k+1)p} P(X = k),$$

for $k-1, k-2, \ldots, 0$ or until convergence. Sum of these probabilities plus $P(X = k)$ is $P(X \le k)$;
else compute $P(X = k + 1)$ using the forward recursion relation

$$P(X = k+1) = \frac{(n-k)p}{(k+1)(1-p)} P(X = k),$$

for $k+1, \ldots, n$, or until convergence; sum these probabilities to get $P(X \ge k+1)$. The cumulative probability

$$P(X \le k) = 1.0 - P(X \ge k+1).$$

The following algorithm for computing the binomial cdf is based on the above method.

Algorithm 3.10.1

```
Input:
      k = nonnegative integer (0 <= k <= n)
      p = success probability (0 < p < 1)
      n = number of trials, n >= 1

Output: bincdf = P(X <= k)

Set  mode = int(n*p)
     bincdf = 0.0d0
     pk = P(X = k)
     if(k .le. mode) then
        For i = k to 0
        bincdf = bincdf + pk
        pk = pk * i*(1.0d0-p)/(en-i+1.0d0)/p
        (end i loop)
     else
        For i = k to n
        pk = pk * (en-i)*p/(i+1.0d0)/(1.0d0-p)
        bincdf = bincdf + pk
        (end i loop)
        bincdf = 1.0d0-bincdf+pk
     end if
```

The following Fortran function routine computes the cdf and pmf of a binomial(n, p) distribution.

```
Input:
      k = the value at which the cdf is to be evaluated,
          k = 0, 1, 2, ... , n
      n = number of trials
      p = success probability

Output:
      P(X <= x) = bincdf(k, n, p)
cccccccccccccccccccccccccccccccccccccccccccccccccccccc
       double precision function bincdf(k, n, p)
       implicit doubleprecision (a-h,o-z)
       ek = k; en = n
```

```
      mode = int(n*p)
      bincdf = 0.0d0; pk = binprob(k, n, p)

      if(k .le. mode) then
         do i = k, 0, -1
            bincdf = bincdf + pk;
            pk = i*(1.0d0-p)*pk/(en-i+1.0d0)/p
         end do
      else
         do i = k, n
            pk = pk * (en-i)*p/(i+1.0d0)/(1.0d0-p)
            bincdf = bincdf + pk
         end do
         bincdf = 1.0d0-bincdf+pk
      end if
      end

ccccccccccccccccccccccccccccccccccccccccccccccccccccccccc
      double precision function binprob(k, n, p)
      implicit doubleprecision (a-h,o-z)
      ek = k; en = n
c
c alng(x) = logarithmic function given in Section 1.8
c
      term = alng(en+1.0d0) - alng(ek+1.0d0) - alng(en-ek+1.0d0)
     &     + ek*dlog(p) + (en-ek)*dlog(1.0d0-p)
     &
      binprob = dexp(term)
      end
```

Chapter 4

Hypergeometric Distribution

4.1 Description

Consider a lot consisting of N items of which M of them are defective and the remaining $N - M$ of them are nondefective. A sample of n items is drawn randomly without replacement. (That is, an item sampled is not replaced before selecting another item.) Let X denote the number of defective items that is observed in the sample. The random variable X is referred to as the hypergeometric random variable with parameters N and M. Noting that the number of ways one can select b different objects from a collection of a different objects is

$$\binom{a}{b} = \frac{a!}{b!(a-b)!},$$

we find that the number of ways of selecting k defective items from M defective items is $\binom{M}{k}$; the number of ways of selecting $n - k$ nondefective items from $N-M$ nondefective items is $\binom{N-M}{n-k}$. Therefore, total number of ways of selecting n items with k defective and $n-k$ nondefective items is $\binom{M}{k}\binom{N-M}{n-k}$. Finally, the number of ways one can select n different items from a collection of N different items is $\binom{N}{n}$. Thus, the probability of observing k defective items in a sample of n items is given by

$$f(k|n, M, N) = P(X = k|n, M, N) = \frac{\binom{M}{k}\binom{N-M}{n-k}}{\binom{N}{n}}, \quad L \le k \le U,$$

where $L = \max\{0, M - N + n\}$ and $U = \min\{n, M\}$.

The cumulative distribution function of X is given by

$$F(k|n, M, N) = \sum_{i=L}^{k} \frac{\binom{M}{i}\binom{N-M}{n-i}}{\binom{N}{n}}, \quad L = \max\{0,\ M - N + n\}.$$

We shall denote the distribution by hypergeometric(n, M, N). The plots of probability mass functions are given in Figure 4.1 for a small lot size of $N = 100$ (the first set of four plots) and for a large lot size of $N = 5,000$ (the second set of eight plots). The parameter-sample size configurations are chosen so that the hypergeometric plots can be compared with the corresponding binomial plots in Figure 3.1. The binomial plots with $n = 20$ are not in good agreement with the corresponding hypergeometric plots in Figure 4.1 with $(N = 100, n = 20)$ whereas all binomial plots are almost identical with the hypergeometric plots with $(N = 5000, n = 20)$ and with $(N = 5000, n = 100)$. Also, see Burstein (1975).

4.2 Moments

Mean:	$n\left(\frac{M}{N}\right)$
Variance:	$n\left(\frac{M}{N}\right)\left(1 - \frac{M}{N}\right)\left(\frac{N-n}{N-1}\right)$
Mode:	The largest integer $\le \frac{(n+1)(M+1)}{N+2}$
Mean Deviation:	$\frac{2x(N-M-n+x)\binom{M}{x}\binom{N-M}{n-x}}{N\binom{N}{n}}$, where x is the smallest integer larger than the mean. [Kamat (1965)]
Coefficient of Variation:	$\left(\frac{(N-M)(N-n)}{nM(N-1)}\right)^{1/2}$
Coefficient of Skewness:	$\frac{(N-2M)(N-2n)\sqrt{(N-1)}}{(N-2)\sqrt{nM(N-M)(N-n)}}$
Coefficient of Kurtosis:	$\left(\frac{N^2(N-1)}{nM(N-M)(N-2)(N-3)(N-n)}\right) \times \left(\frac{3nM(N-M)(6-n)}{N} + N(N+1-6n) + 6n^2 + 3M(N-M)(n-2) - \frac{18n^2M(N-M)}{N^2}\right)$

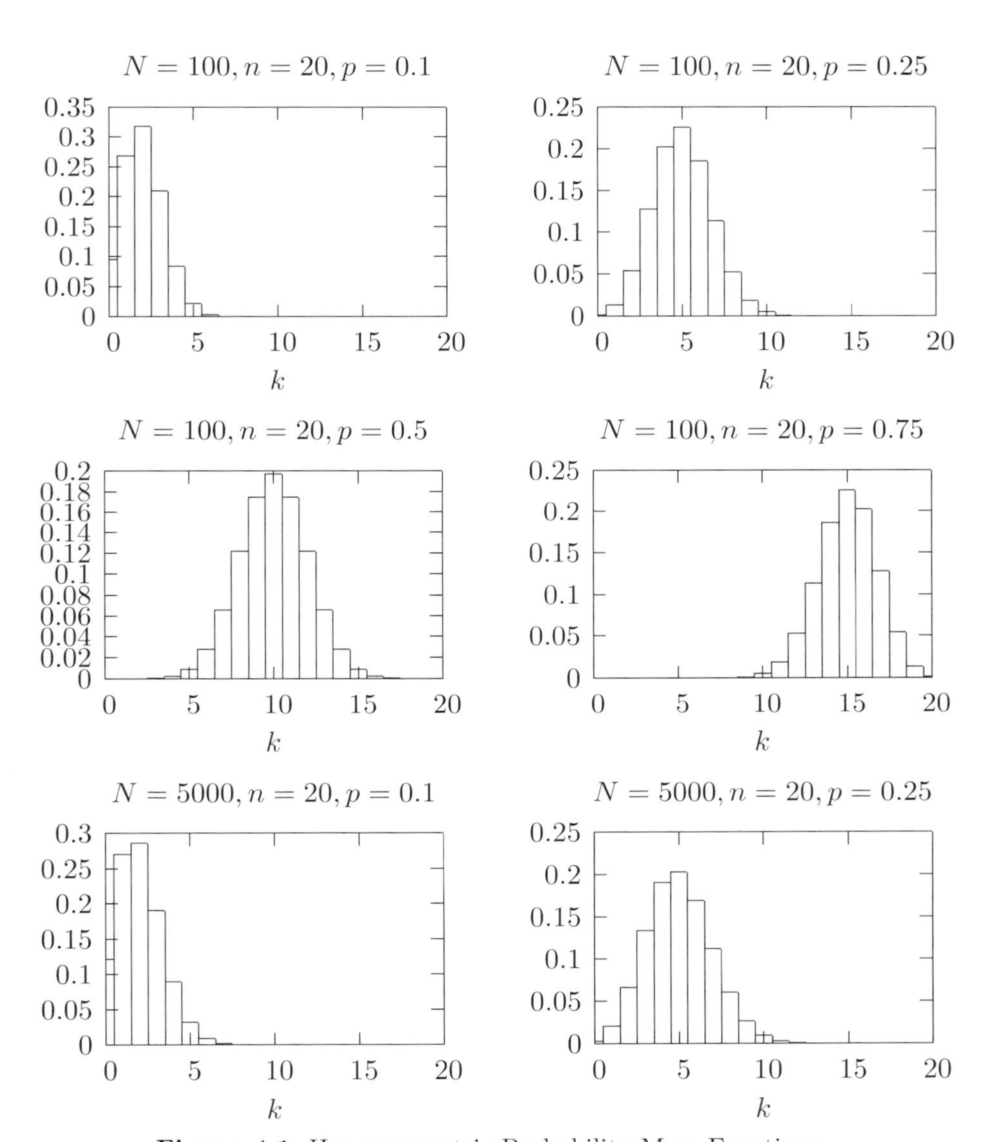

Figure 4.1 Hypergeometric Probability Mass Functions

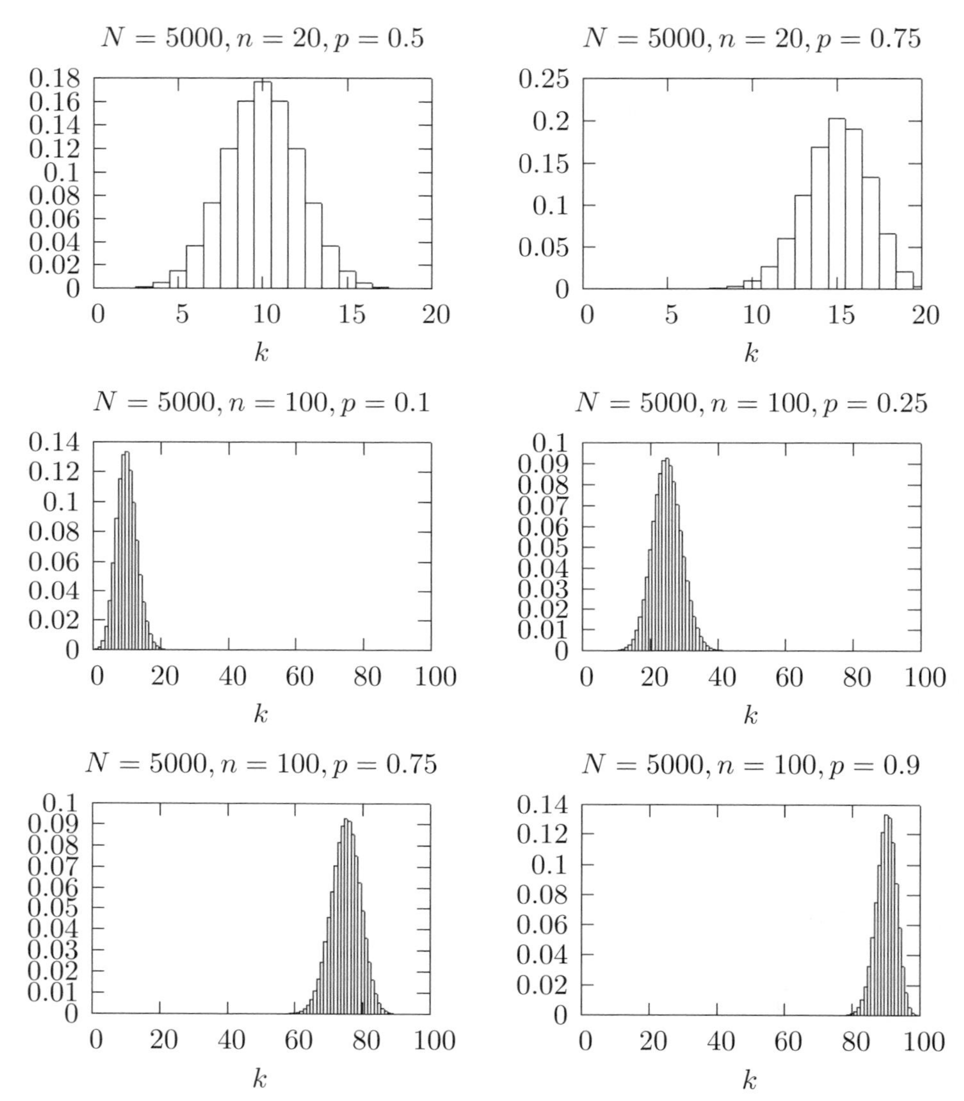

Figure 4.1 Hypergeometric Probability Mass Functions (continued)

4.3 Computing Table Values

The dialog box [StatCalc→Discrete→Hypergeometric] computes the following values.

1. Tail Probabilities, Critical Values and Moments.
2. Test for Proportion and Power Calculation [Section 4.5].

3. Confidence Interval for Proportion and Sample Size for Precision [Section 4.6].
4. Test for the Difference between Two Proportions and Power Calculation [Section 4.7].

The dialog box [StatCalc→Discrete→Hypergeometric →Probabilities, Critical Values and Moments] computes the probabilities, moments and other parameters of a hypergeometric(n, M, N) distribution.

Example 4.3.1 When $N = 100$, $M = 36$, $n = 20$, and $k = 3$,

$$P(X \leq 3) = 0.023231, \quad P(X \geq 3) = 0.995144 \text{ and } P(X = 3) = 0.018375.$$

To Compute other Parameters: For any given four values of $\{N, M, n, k, P(X \leq k)\}$, *StatCalc* computes the missing one. For example, to compute the value of M for a given N, n, k and $P(X \leq k)$, enter the values of N, n, k and $P(X \leq k)$, and then click on [E].

Example 4.3.2 When $N = 300$, $n = 45$, $k = 12$, and $P(X \leq k) = 0.4321$, the value of M is 87. To carry out the computation, enter 300 for N, 45 for n, 12 for k and 0.4321 for $P(X \leq k)$, and then click [E]. Note that $P(X \leq 12|45, 87, 300) = 0.429135$, which is close to the specified probability 0.4321.

The values of N, n, and k can also be similarly calculated.

To compute moments: Enter the values of the N, M, and n; click [M].

Illustrative Examples

Example 4.3.3 The following state lottery is well-known in the United States of America. A player selects 6 different numbers from $1, 2, \ldots, 44$ by buying a ticket for \$1.00. Later in the week, the winning numbers will be drawn randomly by a device. If the player matches all six winning numbers, then he or she will win the jackpot of the week. If the player matches 4 or 5 numbers, he or she will receive a lesser cash prize. If a player buys one ticket, what are the chances of matching

a. all six numbers?

b. four numbers?

Solution: Let X denote the number of winning numbers in the ticket. If we regard winning numbers as defective, then X is a hypergeometric random variable with $N = 44$, $M = 6$, and $n = 6$. The probabilities can be computed using the dialog box [StatCalc→Discrete→Hypergeometric →Probabilities, Critical Values and Moments].

a.

$$P(X=6) = \frac{\binom{6}{6}\binom{38}{0}}{\binom{44}{6}} = \frac{1}{\binom{44}{6}} = \frac{6!\,38!}{44!} = \frac{1}{7059052}.$$

b.

$$P(X=4) = \frac{\binom{6}{4}\binom{38}{2}}{\binom{44}{6}} = 0.0014938.$$

To find the probability in part (b) using *StatCalc*, enter 44 for N, 6 for M, 6 for sample size n, and 4 for observed k, and click [P] to get $\mathrm{P}(X=4) = 0.0014938$.

Example 4.3.4 A shipment of 200 items is under inspection. The shipment will be acceptable if it contains 10 or fewer defective items. The buyer of the shipment decided to buy the lot if he finds no more than one defective item in a random sample of n items from the shipment. Determine the sample size n so that the chances of accepting an unacceptable shipment is less than 10%.

Solution: Since we deal with a finite population, a hypergeometric model with $N = 200$ is appropriate for this problem. Let X denote the number of defective items in a sample of n items. The shipment is unacceptable if the number of defective items M is 11 or more. Furthermore, note that for M in $\{11, 12, \ldots, 200\}$, the chances of accepting an unacceptable shipment, that is, $P(X \le 1|n, M, N)$, attains the maximum when $M = 11$. So we need to determine the value of n so that $P(X \le 1|n, 11, 200) \le 0.10$ and $P(X \le 1|(n-1), 11, 200) > 0.10$. To compute the required sample size using *StatCalc*, select [StatCalc→Discrete→Hypergeometric→Probabilities, Critical Values and Moments], enter 200 for N, 11 for M, 1 for k and 0.1 for $P(X \le k)$; click [S] to get 61. Note that $P(X \le 1|61, 11, 200) = 0.099901$.

Thus, the buyer has to inspect a sample of 61 items so that the chances of accepting an unacceptable shipment is less than 10%. Also, notice that when the sample size is 60, the probability of accepting an unacceptable lot is 0.106241, which is greater than 10%.

4.4 Point Estimation

Let k denote the observed number of defective items in a sample of n items, selected from a lot of N items. Let M denote the number of defective items in the lot.

Point Estimation of M

The maximum likelihood estimator of M is given by

$$\widehat{M} = \left[\frac{k(N+1)}{n}\right],$$

where $[x]$ denotes the largest integer less than or equal to x. If $k(N+1)/n$ is an integer, then both $k(N+1)/n$ and $k(N+1)/n - 1$ are the maximum likelihood estimators of M.

Estimation of the Lot Size

There are situations in which we want to estimate the lot size based on M, n, and k. For example, the capture-recapture technique is commonly used to estimate animal abandons in a given region [Thompson (1992), p.212]: A sample of n_1 animals was trapped, marked, and released in the first occurrence. After a while, another sample of n_2 animals was trapped. Let X denote the number of marked animals in the second trap. Then X follows a hypergeometric distribution with $M = n_1$ and $n = n_2$. For given $X = k$, we want to estimate N (lot size = total number of animals in the region). The maximum likelihood estimator of N is given by

$$\widehat{N} = \left[\frac{n_1 n_2}{k}\right],$$

where $[x]$ denotes the largest integer less than or equal to x.

4.5 Test for the Proportion

Suppose that we found k defective items in a sample of n items drawn from a lot of N items. Let M denote the true number of defective items in the population and $p = M/N$.

4.5.1 An Exact Test

Let $M_0 = \text{int}(Np_0)$. For testing

$$H_0 : p \le p_0 \quad \text{vs.} \quad H_a : p > p_0, \tag{4.5.1}$$

the null hypothesis will be rejected if the p-value $P(X \ge k|n, M_0, N) \le \alpha$, for testing

$$H_0 : p \ge p_0 \quad \text{vs.} \quad H_a : p < p_0, \tag{4.5.2}$$

the null hypothesis will be rejected if the p-value $P(X \le k|n, M_0, N) \le \alpha$, and for testing

$$H_0 : p = p_0 \quad \text{vs.} \quad H_a : p \ne p_0, \tag{4.5.3}$$

the null hypothesis will be rejected if the p-value

$$2\min\{P(X \le k|n, M_0, N), P(X \ge k|n, M_0, N)\} \le \alpha.$$

4.5.2 Power of the Exact Test

For a given p, let $M = \text{int}(Np)$ and $M_0 = \text{int}(Np_0)$, where p_0 is the specified value of p under H_0 in (4.5.1). For a right-tail test, the exact power can be computed using the expression

$$\sum_{k=L}^{U} \frac{\binom{M}{k}\binom{N-M}{n-k}}{\binom{N}{n}} I(P(X \ge k|n, M_0, N) \le \alpha),$$

and for a two-tail test the exact power can be expressed as

$$\sum_{k=L}^{U} \frac{\binom{M}{k}\binom{N-M}{n-k}}{\binom{N}{n}} I(P(X \ge k|n, M_0, N) \le \alpha/2 \text{ or } P(X \le k|n, M_0, N) \le \alpha/2),$$

where $I(.)$ is the indicator function.

The dialog box [StatCalc→Discrete→Hypergeometric→Test for p and Power Calculation] uses the above formulas for computing p-values and powers.

Example 4.5.1 (*Calculation of p-values*) When $N = 500$, $n = 20$, $k = 8$, and $p_0 = 0.2$, it is desired to test $H_0 : p \le p_0$ vs. $H_a : p > p_0$ at the level of 0.05. After entering these values in the dialog box, click [p-values for] to get 0.0293035. If the nominal level is 0.05, then the null hypothesis H_0 will be rejected in favor of the alternative hypothesis H_a. However, if $H_0 : p = p_0$ vs. $H_a : p \ne p_0$ then, at the same nominal level, the H_0 in (4.5.3) can not be rejected because the p-value for this two-tail test is 0.058607, which is not less than 0.05.

Example 4.5.2 (*Power Calculation*) When $N = 500$, Sample Size = 35, $p_0 = 0.2$, nominal level = 0.05, and $p = 0.4$, the power of the test for the hypotheses in (4.5.1) is 0.813779. For hypotheses in (4.5.3), the power is 0.701371. The power can be computed using *StatCalc* as follows. Enter 500 for N, 0.2 for [Value of p0], 3 for [two-tail], .05 for [Level], 0.4 for [Guess p] and 35 for [S Size]; click on [Power].

Example 4.5.3 (*Sample Size Calculation*) Suppose that a researcher believes that a new drug is 20% more effective than the existing drug which has a success rate of 70%. Assume that the size of the population of patients is 5000. The required sample size to test his belief (at the level 0.05 and power 0.90) can be computed using *StatCalc* as follows. Enter 5000 for N, 0.70 for p_0, 1 for the right-tail test, 0.05 for the level, 0.90 for [Guess p], 0.90 for [Power], and click on [S Size] to get 37. To get the actual power, click on [Power] to get 0.929651.

Example 4.5.4 A pharmaceutical company claims that 75% of doctors prescribe one of its drugs for a particular disease. In a random sample of 40 doctors from a population of 1000 doctors, 23 prescribed the drug to their patients. Does this information provide sufficient evidence to indicate that the true percentage of doctors who prescribe the drug is less than 75? Test at the level of significance $\alpha = 0.05$.

Solution: Let p denote the actual proportion of doctors who prescribe the drug to their patients. We want to test

$$H_0 : p \geq 0.75 \qquad \text{vs.} \qquad H_a : p < 0.75.$$

In the dialog box [StatCalc→Discrete→Hypergeometric→Test for p and Power Calculation], enter 1000 for N, 40 for n, 23 for observed k and 0.75 for [Value of p0], and click on [p-values for]. The p-value for the above left-tail test is 0.0101239, which is less than 0.05. Thus, we conclude, on the contrary to the manufacturer's claim, that 75% or less doctors prescribe the drug.

4.6 Confidence Interval and Sample Size Calculation

Suppose that we found k defective items in a sample of n items drawn from a finite population of N items. Let M denote the true number of defective items in the population.

4.6.1 Confidence Intervals

A lower confidence limit M_L for M is the largest integer such that

$$P(X \geq k|n, M_L, N) \leq \alpha/2,$$

and an upper limit M_U for M is the smallest integer such that

$$P(X \leq k|n, M_U, N) \leq \alpha/2.$$

A $1-\alpha$ confidence interval for the proportion of defective items in the lot is given by

$$(p_L,\ p_U) = \left(\frac{M_L}{N},\ \frac{M_U}{N}\right). \tag{4.5.4}$$

The dialog box [StatCalc→Discrete→Hypergeometric→CI for p and Sample size for Precision] uses the above formulas to compute $1-\alpha$ confidence intervals for p. This dialog box also computes the sample size required to estimate the proportion of defective items within a given precision.

4.6.2 Sample Size for Precision

For a given N, n, and k, the dialog box [StatCalc→Discrete→Hypergeometric →CI for p and Sample Size for Precision] computes the one-sided confidence limits and confidence interval for p.

Expected Length

For a given a lot size N, p and a confidence level $1-\alpha$, the expected length of the confidence interval in (4.5.4) can be computed as follows. For a given p, let $M = \text{int}(Np)$. Then the expected length is given by

$$\sum_{k=L}^{U} \frac{\binom{M}{k}\binom{N-M}{n-k}}{\binom{N}{n}}(p_U - p_L).$$

StatCalc computes the required sample size to have a confidence interval with the desired expected length.

Example 4.6.1 (*Computing Confidence Interval*) Suppose that a sample of 40 units from a population of 500 items showed that 5 items are with an attribute of interest. To find a 95% confidence interval for the true proportion of the items with this attribute, enter 500 for N, 40 for n, 5 for the observed number of successes k, and 0.95 for the confidence level; click [2-sided] to get (0.042, 0.262). For one-sided limits, click [1-sided] to get 0.05 and 0.24. That is, 0.05 is 95% one-sided lower limit for p, and 0.24 is 95% one-sided upper limit for p.

Example 4.6.2 (*Sample Size for a Given Precision*) A researcher hypothesizes that the proportion of individuals with the attribute of interest in a population of size 1000 is 0.3, and he wants to estimate the true proportion within ±5% with 95% confidence. To compute the required sample size, enter 1000 for [Lot Size], 0.95 for [Conf Level], 0.3 for [Guess p] and 0.05 for [Half-Width]. Click

[Sam Size] to get 282. See Example 3.5.3 to find out the required sample size if the population is infinite or its size is unknown.

Example 4.6.3 (*Confidence Interval*) A shipment of 320 items is submitted for inspection. A random sample of 50 items was inspected, and it was found that 5 of them were defective. Find a 95% confidence interval for the number of defective items in the lot.

Solution: Select the dialog box [StatCalc→Discrete→Hypergeometric → CI for p and Sample Size for Precision], enter 320 for the lot size N, 50 for the sample size n, and 5 for the observed k; click [2-sided] to get (0.034375, 0.209375). That is, the actual percentage of defective items is between 3.4% and 20.9% with 95% confidence. To get the confidence interval for M, multiply both endpoints by $N = 320$; this gives (11, 67).

To get 95% one-sided limits, click on [1-sided] to get 0.040625 and 0.19375. This means that the true proportion of defective items is at least 0.040625. The 95% one-sided upper limit is 0.19375.

Example 4.6.4 (*One-Sided Limit*) A highway patrol officer stopped a car for a minor traffic violation. Upon suspicion, the officer checked the trunk of the car, and found many bags. The officer arbitrarily checked 10 bags and found that all of them contain marijuana. A later count showed that there were 300 bags. Before the case went to trial, all the bags were destroyed without examining the remaining bags. Since the severity of the fine and punishment depends on the quantity of marijuana, it is desired to estimate the minimum number of marijuana bags. Based on the information, determine the minimum number marijuana bags in the trunk at the time of arrest.

Solution: The hypergeometric model, with lot size $N = 300$, sample size $n = 10$, and the observed number of defective items $k = 10$, is appropriate for this problem. So, we can use a 95% one-sided lower limit for M (total number of marijuana bags) as an estimate for the minimum number of marijuana bags. To get a 95% one-sided lower limit for M using *StatCalc*, enter 300 for N, 10 for n, 10 for k, and 0.95 for the confidence level; click [1-sided] to get 0.743. That is, we estimate with 95% confidence that there were at least 223 (300×0.743) bags of marijuana at the time of arrest.

4.7 A Test for the Difference between Two Proportions

Suppose that inspection of a sample of n_1 individuals from a population of N_1 units revealed k_1 units with a particular attribute, and a sample of n_2 individuals from another population of N_2 units revealed k_2 units with the same attribute. The problem of interest is to test the difference $p_1 - p_2$, where p_i denotes the true proportion of individuals in the ith population with the attribute of interest, $i = 1, 2$.

4.7.1 The Test

Consider testing

$$H_0 : p_1 \le p_2 \quad \text{vs.} \quad H_a : p_1 > p_2. \tag{4.7.1}$$

Define $\widehat{p}_k = \frac{k_1+k_2}{n_1+n_2}$ and $\widehat{M}_i = \text{int}(N_i\widehat{p}_k)$, $i = 1, 2$. The p-value for testing the above hypotheses can be computed using the expression

$$\begin{aligned} P(k_1, k_2, n_1, n_2) &= \sum_{x_1=L_1}^{U_1} \sum_{x_2=L_2}^{U_2} f(x_1|n_1, \widehat{M}_1, N_1) f(x_2|n_2, \widehat{M}_2, N_2) \\ &\times \ I(Z(x_1, x_2) \ge Z(k_1, k_2)), \end{aligned} \tag{4.7.2}$$

where $I(.)$ is the indicator function, $L_i = \max\{0, \widehat{M}_i - N_i + n_i\}$, $U_i = \min\{\widehat{M}_i, n_i\}$, $i = 1, 2$,

$$f(x_i|n_i, \widehat{M}_i, N_i) = \frac{\binom{\widehat{M}_i}{x_i}\binom{N_i-\widehat{M}_i}{n_i-x_i}}{\binom{N_i}{n_i}}, \quad i = 1, 2,$$

$$Z(x_1, x_2) = \frac{x_1 - x_2}{\sqrt{\widehat{p_x}(1-\widehat{p_x})\left(\frac{N_1-n_1}{n_1(N_1-1)} + \frac{N_2-n_2}{n_2(N_2-1)}\right)}}, \quad \text{and} \quad \widehat{p}_x = \frac{x_1 + x_2}{n_1 + n_2}.$$

The term $Z(k_1, k_2)$ is equal to $Z(x_1, x_2)$ with x replaced by k.

The null hypothesis in (4.7.1) will be rejected when the p-value in (4.7.2) is less than or equal to α. For more details see Krishnamoorthy and Thomson (2002). The p-value for a left-tail test or for a two-tail test can be computed similarly.

4.7.2 Power Calculation

For a given p_1 and p_2, let $M_i = \text{int}(N_i p_i)$, $L_i = \max\{0, M_i - N_i + n_i\}$ and $U_i = \min\{n_i, M_i\}$, $i = 1, 2$. The exact power of the test described above can be computed using the expression

$$\sum_{k_1=0}^{n_1} \sum_{k_2=0}^{n_2} f(k_1|n_1, M_1, N_1) f(k_2|n_2, M_2, N_2) I(P(k_1, k_2, n_1, n_2) \leq \alpha), \quad (4.7.3)$$

where $f(k|n, M, N)$ is the hypergeometric pmf, $M_i = \text{int}(N_i p_i)$, $i = 1, 2$, and the p-value $P(k_1, k_2, n_1, n_2)$ is given in (4.7.2). The powers of a left-tail test and a two-tail test can be computed similarly.

The dialog box [StatCalc→Discrete→Hypergeometric→Test for p1-p2 and Power Calculation] uses the above methods for computing the p-values and powers of the above two-sample test.

Example 4.7.1 (Calculation of p-values) Suppose a sample of 25 observations from population 1 with size 300 yielded 20 successes, and a sample of 20 observations from population 2 with size 350 yielded 10 successes. Let p_1 and p_2 denote the proportions of successes in populations 1 and 2, respectively. Suppose we want to test

$$H_0 : p_1 \leq p_2 \quad \text{vs.} \quad H_a : p_1 > p_2.$$

To compute the p-value, enter the numbers of successes and sample sizes, click on [p-values for] to get the p-value of 0.0165608. The p-value for testing $H_0 : p_1 = p_2$ vs. $H_a : p_1 \neq p_2$ is 0.0298794.

Example 4.7.2 (Sample Size Calculation for Power) Suppose the sample size for each group needs to be determined to carry out a two-tail test at the level of significance $\alpha = 0.05$ and power $= 0.80$. Assume that the lot sizes are 300 and 350. Furthermore, the guess values of the proportions are given as $p_1 = 0.45$ and $p_2 = 0.15$. To determine the sample size using *StatCalc*, enter 2 for two-tail test, 0.05 for [Level], 0.45 for p_1, 0.15 for p_2, and 28 for each sample size. Click [Power] to get a power of 0.751881. Note that the sample size gives a power less than 0.80. This means, the sample size required to have a power of 0.80 is more than 28. Enter 31 (for example) for both sample sizes and click on [Power] radio button. Now the power is 0.807982. The power at 30 is 0.78988. Thus, the required sample size from each population to attain a power of at least 0.80 is 31.

Remark 4.7.1 Note that the power can also be computed for unequal sample sizes. For instance, when $n_1 = 30$, $n_2 = 34$, $p_1 = 0.45$, $p_2 = 0.15$, the power for testing $H_0 : p_1 = p_2$ vs. $H_a : p_1 \neq p_2$ at the nominal 0.05 is 0.804974. For the same configuration, a power of 0.800876 can be attained if $n_1 = 29$ and $n_2 = 39$.

4.8 Properties and Results

4.8.1 Recurrence Relations

a. $P(X = k+1|n, M, N) = \frac{(n-k)(M-k)}{(k+1)(N-M-n+k+1)} P(X = k|n, M, N)$.

b. $P(X = k-1|n, M, N) = \frac{k(N-M-n+k)}{(n-k+1)(M-k+1)} P(X = k|n, M, N)$.

c. $P(X = k|n+1, M, N) = \frac{(N-M-n+k)}{(M+1-k)(N-M)} P(X = k|n, M, N)$.

4.8.2 Relation to Other Distributions

1. Binomial: Let X and Y be independent binomial random variables with common success probability p and numbers of trials m and n, respectively. Then

$$P(X = k|X+Y = s) = \frac{P(X = k)P(Y = s-k)}{P(X+Y = s)}$$

which simplifies to

$$P(X = k|X+Y = s) = \frac{\binom{m}{k}\binom{n}{s-k}}{\binom{m+n}{s}}, \quad \max\{0, s-n\} \leq k \leq \min(m, s).$$

Thus, the conditional distribution of X given $X + Y = s$ is hypergeometric$(s, m, m+n)$.

4.8.3 Approximations

1. Let $p = M/N$. Then, for large N and M,

$$P(X = k) \simeq \binom{n}{k} p^k (1-p)^{n-k}.$$

2. Let (M/N) be small and n is large such that $n(M/N) = \lambda$.

$$P(X = k) \simeq \frac{e^{-\lambda}\lambda^k}{k!}\left\{1 + \left(\frac{1}{2M} + \frac{1}{2n}\right)\left[k - \left(k - \frac{Mn}{N}\right)^2\right] + O\left(\frac{1}{k^2} + \frac{1}{n^2}\right)\right\}$$

[Burr (1973)]

4.9 Random Number Generation

```
Input:
      N = lot size; M = number of defective items in the lot
      n = sample size; ns = number of random variates to be
          generated

Output:
      x(1),..., x(ns) are random number from the
      hypergeometric(n, M, N) distribution
```

The following generating scheme is essentially based on the probability mechanism involved in simple random sampling without replacement, and is similar to Algorithm 3.9.1 for the binomial case.

Algorithm 4.9.1

```
Set k = int((n + 1)*(M + 1)/(N + 2))
    pk = P(X = k)
    df = P(X <= k)
    Low = max{0, M - N + n}
    High = min{n, M}
    rpk = pk; rk = k
For j = 1 to ns
    Generate u from uniform(0, 1)
    If u > df, go to 2
1   u = u + pk
    If k = Low or u > df, go to 3
    pk = pk*k*(N - M - n + k)/((M - k + 1)*(n - k + 1))
    k = k - 1
    go to 1
2   pk = pk*(n - k)*(M -k)/((k + 1)*(N - M + k + 1))
    u = u - pk
    k = k + 1
    If k = High or u <= df, go to 3
    go to 2
3   x(j) = k
    pk = rpk
    k = rk
[end j loop]
```

For other lengthy but more efficient algorithms see Kachitvichyanukul and Schmeis (1985).

4.10 Computation of Probabilities

To compute $P(X = k)$

Set $U = \min\{n, M\}$; $L = \max\{0, M - N + n\}$
If $k > U$ or $k < L$ then return $P(X = k) = 0$
Compute $S_1 = \ln\Gamma(M+1) - \ln\Gamma(k+1) - \ln\Gamma(M-k+1)$
$S_2 = \ln\Gamma(N-M+1) - \ln\Gamma(n-k+1) - \ln\Gamma(N-M-n+k+1)$
$S_3 = \ln\Gamma(N+1) - \ln\Gamma(n+1) - \ln\Gamma(N-n+1)$
$P(X = k) = \exp(S_1 + S_2 - S_3)$

To compute $\ln\Gamma(x)$, see Section 1.8.

To compute $P(X \le k)$

Compute $P(X = k)$
Set mode $= \text{int}((n+1)(M+1)/(N+2))$
If $k \le$ mode, compute the probabilities using the backward recursion relation

$$P(X = k-1|n, M, N) = \frac{k(N-M-n+k)}{(n-k+1)(M-k+1)} P(X = k|n, M, N)$$

for $k-1, \ldots, L$ or until a specified accuracy; add these probabilities and $P(X = k)$ to get $P(X \le k)$;
else compute the probabilities using the forward recursion

$$P(X = k+1|n, M, N) = \frac{(n-k)(M-k)}{(k+1)(N-M-n+k+1)} P(X = k|n, M, N)$$

for $k+1, \ldots, U$ or until a specified accuracy; add these probabilities to get $P(X \ge k+1)$. The cumulative probability is given by

$$P(X \le k) = 1 - P(X \ge k+1).$$

The following algorithm for computing a hypergeometric cdf is based on the above computational method.

Algorithm 4.10.1

```
Input:
      k = the value at which the cdf is to be evaluated
      n = the sample size
      m = the number of defective items in the lot
      lot = size of the lot
Output:
      hypcdf = P(X <= k|n,m,lot)

Set   one = 1.0d0
      lup = min(n, m)
      low = max(0, m-lot+n)

      if(k .lt. low) return hypcdf = 0.0d0
      if(k .gt. lup) return hypcdf = one

      mode = int(n*m/lot)
      hypcdf = 0.0d0
      pk = hypprob(k, n, m, lot)

      if(k .le. mode) then
         For i = k to low
           hypcdf = hypcdf + pk
           pk = pk*i*(lot-m-n+i)/(n-i+one)/(m-i+one)
         [end i loop]
      else
         For i = k to lup
           pk = pk * (n-i)*(m-i)/(i+one)/(lot-m-n+i+one)
           hypcdf = hypcdf + pk
         [end i loop]
         hypcdf = 1.0d0-hypcdf
      end if
```

The following Fortran function routines computes the cdf and pmf of a hypergeometric(n, m, lot) distribution.

```
Input:
      k = the value at which the cdf is to be evaluated
      n = sample size
```

```
      m = number of defective items in the lot
    lot = lot size


Output:
      P(X <= x) = hypcdf(k, n, m, lot)

cccccccccccccccccccccccccccccccccccccccccccccccccccccccccccccccccc
      double precision function hypcdf(k, n, m, lot)
      implicit doubleprecision (a-h,o-z)

      lup = min(n, m); low = max(0, m-lot+n)

      one = 1.0d0
      hypcdf = one
      if(k .gt. lup) return

      hypcdf = 0.0d0
      if(k .lt. low) return

      mode = int(n*m/lot)

      hypcdf = 0.0d0; pk = hypprob(k, n, m, lot)

      if(k .le. mode) then
         do i = k, low, -1
            hypcdf = hypcdf + pk;
            pk = pk*i*(lot-m-n+i)/(n-i+one)/(m-i+one)
         end do
      else
         do i = k, lup
            pk = pk * (n-i)*(m-i)/(i+one)/(lot-m-n+i+one)
            hypcdf = hypcdf + pk
         end do
         hypcdf = 1.0d0-hypcdf
      end if

      end
```

```
ccccccccccccccccccccccccccccccccccccccccccccccccccccccccccccc
      double precision function hypprob(k, n, m, lot)
      implicit doubleprecision (a-h,o-z)

      one = 1.0d0
      lup = min(n, m); low = max(0, m-lot+n)

      hypprob = 0.0d0
      if(k .lt. low .or. k .gt. lup) return
c
c alng(x) = logarithmic function given in Section 1.8
c
      term1 = alng(m+one)-alng(k+one)-alng(m-k+one)
      term2 = alng(lot-m+one)-alng(n-k+one)-alng(lot-m-n+k+one)
      term3 = alng(lot+one)-alng(n+one)-alng(lot-n+one)
      hypprob = dexp(term1+term2-term3)

      end
```

Chapter 5

Poisson Distribution

5.1 Description

Suppose that events that occur over a period of time or space satisfy the following:

1. The numbers of events occurring in disjoint intervals of time are independent.
2. The probability that exactly one event occurs in a small interval of time Δ is $\Delta\lambda$, where $\lambda > 0$.
3. It is almost unlikely that two or more events occur in a sufficiently small interval of time.
4. The probability of observing a certain number of events in a time interval Δ depends only on the length of Δ and not on the beginning of the time interval.

Let X denote the number of events in a unit interval of time or in a unit distance. Then, X is called the Poisson random variable with mean number of events λ in a unit interval of time. The probability mass function of a Poisson distribution with mean λ is given by

$$f(k|\lambda) = P(X = k|\lambda) = \frac{e^{-\lambda}\lambda^{k}}{k!}, \quad k = 0, 1, 2, \ldots \tag{5.1.1}$$

The cumulative distribution function of X is given by

$$F(k|\lambda) = P(X \le k|\lambda) = \sum_{i=0}^{k} \frac{e^{-\lambda}\lambda^{i}}{i!}, \quad k = 0, 1, 2, \ldots \tag{5.1.2}$$

The Poisson distribution can also be developed as a limiting distribution of the binomial, in which $n \to \infty$ and $p \to 0$ so that np remains a constant. In other words, for large n and small p, the binomial distribution can be approximated by the Poisson distribution with mean $\lambda = np$. Some examples of the Poisson random variable are:

1. the number of radioactive decays over a period of time;
2. the number of automobile accidents per day on an interstate road;
3. the number of typographical errors per page in a book;
4. the number of α particles emitted by a radioactive source in a unit of time;
5. the number of still births per week in a large hospital.

Poisson distribution gives probability of observing k events in a given period of time assuming that events occur independently at a constant rate. The Poisson distribution is widely used in quality control, reliability, and queuing theory. It can be used to model the distribution of number of defects in a piece of material, customer arrivals at a train station, auto insurance claims, and incoming telephone calls per period of time.

As shown in the plots of probability mass functions in Figure 5.1, Poisson distribution is right-skewed, and the degree of skewness decreases as λ increases.

5.2 Moments

Mean:	λ
Variance:	λ
Mode:	The largest integer less than or equal to λ. If λ is an integer, λ and $\lambda - 1$ are modes.
Mean Deviation:	$\frac{2e^{-\lambda}\lambda^{[\lambda]+1}}{[\lambda]!}$ where $[x]$ denotes the largest integer less than or equal to x. [Johnson, et al. (1992), p. 157]
Coefficient of Variation:	$\frac{1}{\sqrt{\lambda}}$
Coefficient of Skewness:	$\frac{1}{\sqrt{\lambda}}$

Coefficient of Kurtosis:	$3 + \frac{1}{\lambda}$
Factorial Moments:	$E\left(\prod_{i=1}^{k}(X-i+1)\right) = \lambda^k$ $E\left(\prod_{i=1}^{k}(X+i)\right)^{-1} = \frac{1}{\lambda^k}\left(1 - e^{-\lambda}\sum_{i=0}^{k-1}\frac{\lambda^i}{i!}\right)$
Moments about the Mean:	$\mu_k = \lambda\sum_{i=0}^{k-2}\binom{k-1}{i}\mu_i, \quad k = 2, 3, 4, \cdots$ where $\mu_0 = 1$ and $\mu_1 = 0$. [Kendall 1943]
Moment Generating Function:	$\exp[\lambda\,(e^t - 1)]$
Probability Generating Function:	$\exp[\lambda\,(t-1)]$

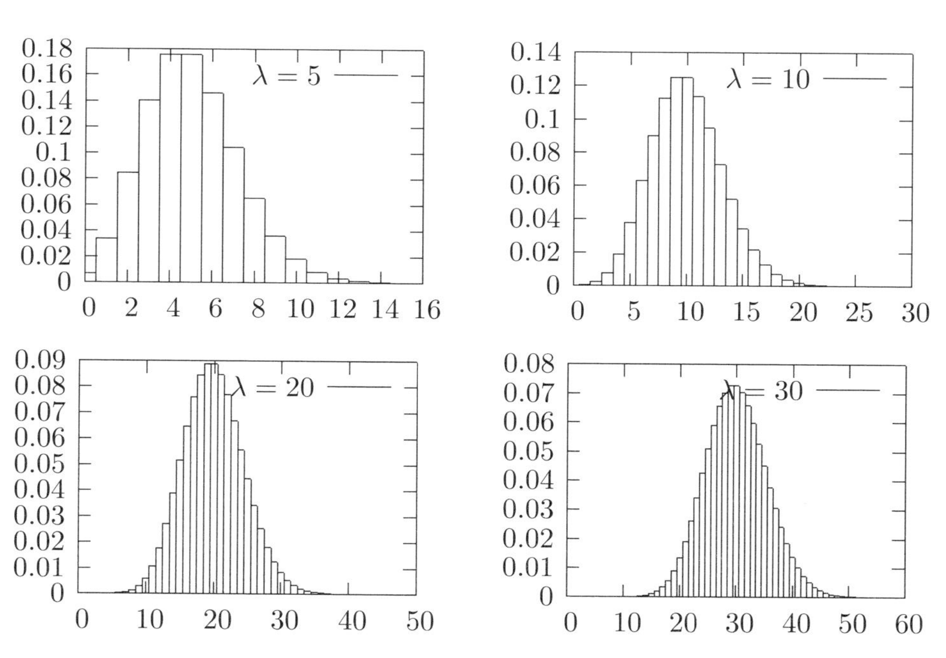

Figure 5.1 Poisson Probability Mass Functions

5.3 Computing Table Values

The dialog box [StatCalc→Discrete→Poisson] computes the following table values, p-values and confidence intervals.

1. Tail Probabilities, Critical Values, and Moments.
2. Test for the Mean and Power Calculation [Section 5.5].
3. Confidence Interval for the Mean and Sample Size for Precision [Section 5.6].
4. Test for the Ratio of Two Means and Power Calculation [Section 5.7].
5. Confidence Interval for the Ratio of Two Means [Section 5.8].
6. An Unconditional Test for the Difference between Two Means and Power Calculation [Section 5.9]

The dialog box [StatCalc→Discrete→Poisson→Probabilities, Critical Values and Moments] computes the tail probabilities, critical points, parameters and moments.

To compute probabilities: Enter the values of the mean, and k at which the probability is to be computed; click [P].

Example 5.3.1 When the mean = 6, $k = 5$, $P(X \leq 5) = 0.44568$, $P(X \geq 5) = 0.714944$ and $P(X = 5) = 0.160623$.

To compute other parameters: *StatCalc* also computes the mean or the value of k when other values are given.

Example 5.3.2 To find the value of the mean when $k = 5$ and $P(X \leq k) = 0.25$, enter 5 for k, enter 0.25 for $P(X \leq k)$, and click [A] to get 7.4227.

Example 5.3.3 To find the value of k, when the mean = 4.5 and $P(X \leq k) = 0.34$, enter these values in *StatCalc*, and click [k] to get 3. Also, note that $P(X \leq 3) = 0.342296$ when the mean is 4.5.

To compute moments: Enter the value of the mean, and click [M].

Example 5.3.4 On average, four customers enter a fast food restaurant per every 3-min period during the peak hours 11:00 am - 1:00 pm. Assuming an approximate Poisson process, what is the probability of 26 or more customers arriving in a 15-min period?

Solution: Let X denote the number of customers entering in a 15-min period. Then, X follows a Poisson distribution with mean = $(4/3) \times 15 = 20$. To find the probability of observing 26 or more customers, select the dialog box [StatCalc→Discrete→Poisson→Probabilities, Critical Values and Moments], enter 20 for the mean, 26 for the observed k and click [P] to get $P(X \geq 26) = 0.1122$.

5.4 Point Estimation

Let $X_1, \ldots, X_n$ be independent observations from a Poisson(λ) population. Then,

$$K = \sum_{i=1}^{n} X_i \sim \text{Poisson}(n\lambda).$$

The following inferences about λ are based on K.

Maximum Likelihood Estimate of λ

The maximum likelihood estimator of λ is given by

$$\widehat{\lambda} = \frac{1}{n}\sum_{i=1}^{n} X_i,$$

which is also the uniformly minimum variance unbiased estimator.

5.5 Test for the Mean

Let $X_1, \ldots, X_n$ be a sample from a Poisson(λ) population. Then,

$$K = \sum_{i=1}^{n} X_i \sim \text{Poisson}(n\lambda).$$

Because the sample size n is known, testing about λ is equivalent to testing about $n\lambda$.

5.5.1 An Exact Test

Let K_0 be an observed value of K. Then, for testing

$$H_0 : \lambda \leq \lambda_0 \ \text{ vs. } \ H_a : \lambda > \lambda_0, \tag{5.5.1}$$

the null hypothesis will be rejected if the p-value $P(K \geq K_0|n\lambda_0) \leq \alpha$, for testing

$$H_0 : \lambda \geq \lambda_0 \quad \text{vs.} \quad H_a : \lambda < \lambda_0, \tag{5.5.2}$$

the null hypothesis will be rejected if the p-value $P(K \leq K_0|n\lambda_0) \leq \alpha$, and for testing

$$H_0 : \lambda = \lambda_0 \quad \text{vs.} \quad H_a : \lambda \neq \lambda_0, \tag{5.5.3}$$

the null hypothesis will be rejected if the p-value

$$2\min\{P(K \leq K_0|n\lambda_0), P(K \geq K_0|n\lambda_0)\} \leq \alpha. \tag{5.5.4}$$

Example 5.5.1 It is desired to find the average number defective spots per 100-ft of an electric cable. Inspection of a sample of twenty 100-ft cable showed an average of 2.7 defective spots. Does this information indicate that the true mean number of defective spots per 100-ft is more than 2? Assuming an approximate Poisson distribution, test using $\alpha = 0.05$.

Solution: Let X denote the number defective spots per 100-f cable. Then, X follows a Poisson(λ) distribution, and we want to test

$$H_0 : \lambda \leq 2 \quad \text{vs.} \quad H_a : \lambda > 2.$$

In the dialog box [StatCalc→Discrete→Poisson→Test for Mean and Power Calculation], enter 20 for the sample size, $20 \times 2.7 = 54$ for the total count, 2 for [Value of M0], and click the [p-values for] to get 0.0199946. Since the p-value is smaller than 0.05, we can conclude that true mean is greater than 2.

5.5.2 Powers of the Exact Test

The exact powers of the tests described in the preceding section can be computed using Poisson probabilities and an indicator function. For example, for a given λ and λ_0, the power of the test for hypotheses in (5.5.1) can be computed using the following expression.

$$\sum_{k=0}^{\infty} \frac{e^{-n\lambda}(n\lambda)^k}{k!} I(P(K \geq k|n\lambda_0) \leq \alpha), \tag{5.5.5}$$

where $K \sim$ Poisson($n\lambda_0$). Powers of the right-tail test and two-tail test can be expressed similarly.

The dialog box [StatCalc→Discrete→Poisson→Test for Mean and Power Calculation] uses the above exact method to compute the power.

Example 5.5.2 (*Sample Size Calculation*) Suppose that a researcher hypothesizes that the mean of a Poisson process has increased from 3 to 4. He likes to determine the required sample size to test his claim at the level 0.05 with power 0.80. To find the sample size, select [StatCalc→Discrete→Poisson→Test for Mean and Power Calculation], enter 3 for [H0: M=M0], 1 for the right-tail test, 0.05 for the level, 4 for [Guess M] and 0.80 for [Power]; click on [S Size] to get 23. To find the actual power at this sample size, click on [Power] to get 0.811302.

5.6 Confidence Intervals for the Mean

Let $X_1, \ldots, X_n$ be a sample from a Poisson(λ) population, and let $K = \sum_{i=1}^{n} X_i$. The following inferences about λ are based on K.

5.6.1 An Exact Confidence Interval

An exact $1-\alpha$ confidence interval for λ is given by (λ_L, λ_U), where λ_L satisfies

$$P(K \geq k|n\lambda_L) = \exp(-n\lambda_L)\sum_{i=k}^{\infty}\frac{(n\lambda_L)^i}{i!} = \frac{\alpha}{2},$$

and λ_U satisfies

$$P(K \leq k|n\lambda_U) = \exp(-n\lambda_U)\sum_{i=0}^{k}\frac{(n\lambda_U)^i}{i!} = \frac{\alpha}{2},$$

where k is an observed value of K. Furthermore, using a relation between the Poisson and chi-square distributions, it can be shown that

$$\lambda_L = \frac{1}{2n}\chi^2_{2k,\alpha/2} \text{ and } \lambda_U = \frac{1}{2n}\chi^2_{2k+2,1-\alpha/2},$$

where $\chi^2_{m,p}$ denotes the pth quantile of a chi-square distribution with df $= m$. These formulas should be used with the convention that $\chi^2_{0,p} = 0$.

The dialog box [StatCalc→Discrete→Poisson→CI for Mean and Sample Size for Width] computes the confidence interval using the above formulas.

Example 5.6.1 (*Confidence Interval for Mean*) Let us compute a 95% confidence interval for the data given in Example 5.5.1. Recall that $n = 20$, sample mean $= 2.7$, and so the total count is 54. To find confidence intervals for the mean number of defective spots, select [StatCalc→Discrete→Poisson→CI for Mean and Sample Size for Width], enter 20 for [Sample Size] 54 for [Total] and 0.95 for [Conf Level]; click [2-sided] to get 2.02832 and 3.52291. That is, (2.03, 3.52) is a 95% confidence interval for the mean. Click [1-sided] to get 2.12537 and 3.387. That is, 2.13 is a 95% one-sided lower limit and 3.39 is a 95% upper limit.

5.6.2 Sample Size Calculation for Precision

For a given n and λ, the expected length of the $1-\alpha$ confidence interval (λ_L, λ_U) in Section 5.6.1 can be expressed as

$$\sum_{k=0}^{\infty} \frac{e^{-n\lambda}(n\lambda)^k}{k!}(\lambda_U - \lambda_L) = \frac{1}{2n}\sum_{k=0}^{\infty} \frac{e^{-n\lambda}(n\lambda)^k}{k!}(\chi^2_{2k+2,1-\alpha/2} - \chi^2_{2k,\alpha/2}).$$

The dialog box [StatCalc→Discrete→Poisson→CI for Mean and Sample Size for Width] also computes the sample size required to estimate the mean within a given expected length.

Example 5.6.2 (Sample Size Calculation) Suppose that a researcher hypothesizes that the mean of a Poisson process is 3. He likes to determine the required sample size to estimate the mean within ± 0.3 and with confidence 0.95. To find the sample size, select [StatCalc→Discrete→Poisson→CI for Mean and Sample Size for Width], enter 0.95 for [Conf Level], 3 for [Guess Mean], and 0.3 for [Half-Width]; click [Sam Size] to get 131.

5.7 Test for the Ratio of Two Means

Let $X_{i1}, \ldots, X_{in_i}$ be a sample from a Poisson(λ_i) population. Then,

$$K_i = \sum_{j=1}^{n_i} X_{ij} \sim \text{Poisson}(n_i\lambda_i), \ i = 1, 2.$$

The following tests about (λ_1/λ_2) are based on the conditional distribution of K_1 given $K_1 + K_2 = m$, which is binomial$(m, n_1\lambda_1/(n_1\lambda_1 + n_2\lambda_2))$.

5.7.1 A Conditional Test

Consider testing

$$H_0 : \frac{\lambda_1}{\lambda_2} \le c \quad \text{vs.} \quad H_a : \frac{\lambda_1}{\lambda_2} > c, \tag{5.7.1}$$

where c is a given positive number. The p-value based on the conditional distribution of K_1 given $K_1 + K_2 = m$ is given by

$$P(K_1 \geq k|m,p) = \sum_{x=k}^{m} \binom{m}{x} p^x (1-p)^{m-x}, \text{ where } p = \frac{n_1 c/n_2}{1+n_1 c/n_2}. \qquad (5.7.2)$$

The conditional test rejects the null hypothesis whenever the p-value is less than or equal to the specified nominal α. [Chapman 1952]

The p-value of a left-tail test or a two-tail test can be expressed similarly.

The dialog box [StatCalc→Discrete→Poisson→Test for Mean1/Mean2 and Power Calculation] uses the above exact approach to compute the p-values of the conditional test for the ratio of two Poisson means.

Example 5.7.1 (*Calculation of p-value*) Suppose that a sample of 20 observations from a Poisson(λ_1) distribution yielded a total of 40 counts, and a sample of 30 observations from a Poisson(λ_2) distribution yielded a total of 22 counts. We like to test

$$H_0 : \frac{\lambda_1}{\lambda_2} \leq 2 \text{ vs. } H_a : \frac{\lambda_1}{\lambda_2} > 2.$$

To compute the p-value using *StatCalc*, enter the sample sizes, total counts, and 2 for the value of c in [H0:M1/M2 = c], and click on [p-values for] to get 0.147879. Thus, there is not enough evidence to indicate that λ_1 is larger than $2\lambda_2$.

Example 5.7.2 (*Calculation of p-values*) Suppose that the number of work related accidents over a period of 12 months in a manufacturing industry (say, A) is 14. In another manufacturing industry B, which is similar to A, the number of work related accidents over a period of 9 months is 8. Assuming that the numbers of accidents in both industries follow Poisson distributions, it is desired to test if the mean number of accidents per month in industry A is greater than that in industry B. That is, we want to test

$$H_0 : \frac{\lambda_1}{\lambda_2} \leq 1 \text{ vs. } H_a : \frac{\lambda_1}{\lambda_2} > 1,$$

where λ_1 and λ_2, respectively, denote the true mean numbers of accidents per month in A and B. To find the p-value using *StatCalc*, select [StatCalc→Discrete→Poisson→Test for Mean1/Mean2 and Power Calculation], enter 12 for [Sam Size 1], 9 for [Sam Size 2], 14 for [No. Events 1], 8 for [No. Events 2], 1 for c in [H0:M1/M2 = c], and click [p-values for] to get 0.348343. Thus, we do not have enough evidence to conclude that $\lambda_1 > \lambda_2$.

5.7.2 Powers of the Conditional Test

For given sample sizes, guess values of the means and a level of significance, the exact power of the conditional test in for (5.7.1) can be calculated using the following expression:

$$\sum_{i=0}^{\infty}\sum_{j=0}^{\infty}\frac{e^{-n_1\lambda_1}(n_1\lambda_1)^i}{i!}\frac{e^{-n_2\lambda_2}(n_2\lambda_2)^j}{j!}I(P(X_1 \geq i|i+j,p) \leq \alpha), \tag{5.7.3}$$

where $P(X_1 \geq k|m,p)$ and p are as defined in (5.7.2). The powers of a two-tail test and left-tail test can be expressed similarly.

The dialog box [StatCalc→Discrete→Poisson→Test for Mean1/Mean2 and Power Calculation] uses (5.7.3) to compute the power of the conditional test for the ratio of two Poisson means.

Example 5.7.3 (Sample Size Calculation) Suppose that a researcher hypothesizes that the mean $\lambda_1 = 3$ of a Poisson population is 1.5 times larger than the mean λ_2 of another population, and he likes to test

$$H_0 : \frac{\lambda_1}{\lambda_2} \leq 1.5 \quad \text{vs.} \quad H_a : \frac{\lambda_1}{\lambda_2} > 1.5.$$

To find the required sample size to get a power of 0.80 at the level 0.05, enter 30 for both sample sizes, 1 for one-tail test, 0.05 for level, 3 for [Guess M1], 2 for [Guess M2] and click power to get 0.76827. By raising the sample size to 33, we get a power of 0.804721. Furthermore, when both sample sizes are 32, the power is 0.793161. Therefore, the required sample size is 33.

We note that *StatCalc* also computes the power for unequal sample sizes. For example, when the first sample size is 27 and the second sample size is 41, the power is 0.803072.

For the above example, if it is desired to find sample sizes for testing the hypotheses

$$H_0 : \frac{\lambda_1}{\lambda_2} = 1.5 \quad \text{vs.} \quad H_a : \frac{\lambda_1}{\lambda_2} \neq 1.5,$$

then enter 2 for two-tail test (while keep the other values as they are), and click [Power]. For example, when both sample sizes are 33, the power is 0.705986; when they are 40, the power is 0.791258, and when they are 41 the power is 0.801372.

5.8 Confidence Intervals for the Ratio of Two Means

The following confidence interval for (λ_1/λ_2) is based on the conditional distribution of K_1 given in (5.7.2). Let

$$p = \frac{n_1\lambda_1}{n_1\lambda_1 + n_2\lambda_2} = \frac{(n_1\lambda_1/n_2\lambda_2)}{(n_1\lambda_1/n_2\lambda_2) + 1}.$$

For given $K_1 = k$ and $K_1 + K_2 = m$, a $1 - \alpha$ confidence interval for λ_1/λ_2 is

$$\left(\frac{n_2 p_L}{n_1(1 - p_L)}, \frac{n_2 p_U}{n_1(1 - p_U)}\right),$$

where (p_L, p_U) is a $1 - \alpha$ confidence interval for p based on k successes from a binomial(m, p) distribution (see Section 3.5). The dialog box [StatCalc→Discrete→ Poisson→CI for Mean1/Mean2] uses the above formula to compute the confidence intervals for the ratio of two Poisson means.

Example 5.8.1 (CI for the Ratio of Means) Suppose that a sample of 20 observations from a Poisson(λ_1) distribution yielded a total of 40 counts, and a sample of 30 observations from a Poisson(λ_2) distribution yielded a total of 22 counts. To compute a 95% confidence interval for the ratio of means, enter the sample sizes, total counts, and 0.95 for confidence level in the appropriate edit boxes, and click on [2-sided] to get (1.5824, 4.81807). To get one-sided confidence intervals click on [1-sided] to get 1.71496 and 4.40773. That is, 95% lower limit for the ratio λ_1/λ_2 is 1.71496 and 95% upper limit for the ratio λ_1/λ_2 is 4.40773.

5.9 A Test for the Difference between Two Means

This test is more powerful than the conditional test given in Section 5.7. However, this test is approximate and in some situations the Type I error rates are slightly more than the nominal level. For more details, see Krishnamoorthy and Thomson (2004).

Let $X_{i1}, \ldots, X_{in_i}$ be a sample from a Poisson(λ_i) distribution, $i = 1, 2$. Then,

$$K_i = \sum_{j=1}^{n_i} X_{ij} \sim \text{Poisson}(n_i\lambda_i), \quad i = 1, 2.$$

The following tests about $\lambda_1 - \lambda_2$ are based on K_1 and K_2.

5.9.1 An Unconditional Test

Consider testing

$$H_0: \lambda_1 - \lambda_2 \le d \quad \text{vs.} \quad H_a: \lambda_1 - \lambda_2 > d, \tag{5.9.1}$$

where d is a specified number. Let (k_1, k_2) be an observed value of (K_1, K_2) and let

$$\widehat{\lambda}_d = \frac{k_1 + k_2}{n_1 + n_2} - \frac{dn_1}{n_1 + n_2}.$$

The p-value for testing (5.9.1) is given by

$$P(k_1, k_2) = \sum_{x_1=0}^{\infty} \sum_{x_2=0}^{\infty} \frac{e^{-\eta}\eta^{x_1}}{x_1!} \frac{e^{-\delta}\delta^{x_2}}{x_2!} I(Z(x_1, x_2) \ge Z(k_1, k_2)), \tag{5.9.2}$$

where $\eta = n_1(\widehat{\lambda}_d + d)$, $\delta = n_2\widehat{\lambda}_d$,

$$Z(x_1, x_2) = \frac{\frac{x_1}{n_1} - \frac{x_2}{n_2} - d}{\sqrt{\frac{x_1}{n_1^2} + \frac{x_2}{n_2^2}}}$$

and $Z(k_1, k_2)$ is $Z(x_1, x_2)$ with x replaced by k. The null hypothesis will be rejected whenever the p-value is less than or equal to the nominal level α.

The dialog box [StatCalc→Discrete→Poisson→Test for Mean1 - Mean2 and Power Calculation] in *StatCalc* uses the above formula to compute the p-values for testing the difference between two means.

Example 5.9.1 (*Unconditional Test*) Suppose that a sample of 20 observations from a Poisson(λ_1) distribution yielded a total of 40 counts, and a sample of 30 observations from a Poisson(λ_2) distribution yielded a total of 22 counts. We like to test

$$H_0: \lambda_1 - \lambda_2 \le 0.7 \quad \text{vs.} \quad H_a: \lambda_1 - \lambda_2 > 0.7.$$

To compute the p-value, enter the sample sizes, total counts, and 0.7 for the value of d in [H0:M1-M2 = d], and click on [p-values for] to get 0.0459181. So, at the 5% level, we can conclude that there is enough evidence to indicate that λ_1 is 0.7 unit larger than λ_2.

Example 5.9.2 (*Unconditional Test*) Let us consider Example 5.7.2, where we used the conditional test for testing $\lambda_1 > \lambda_2$. We shall now apply the unconditional test for testing

$$H_0: \lambda_1 - \lambda_2 \le 0 \quad \text{vs.} \quad H_a: \lambda_1 - \lambda_2 > 0.$$

To find the p-value, enter 12 for the sample size 1, 9 for the sample size 2, 14 for [No. Events 1], 8 for [No. Events 2], 0 for d, and click [p-values for] to get 0.279551. Thus, we do not have enough evidence to conclude that $\lambda_1 > \lambda_2$. Notice that the p-value of the conditional test in Example 5.7.2 is 0.348343.

5.9.2 Powers of the Unconditional Test

For a given λ_1, λ_2, and a level of significance α, the power of the unconditional test is given by

$$\sum_{k_1=0}^{\infty}\sum_{k_2=0}^{\infty} \frac{e^{-n_1\lambda_1}(n_1\lambda_1)^{k_1}}{k_1!}\frac{e^{-n_2\lambda_2}(n_2\lambda_2)^{k_2}}{k_2!} I(P(k_1,k_2) \le \alpha), \tag{5.9.3}$$

where $P(k_1, k_2)$ is the p-value given in (5.9.2). When $\lambda_1 = \lambda_2$, the above formula gives the size (that is, actual Type I error rate) of the test.

The dialog box [StatCalc→Discrete→Poisson→Test for Mean1 - Mean2 and Power Calculation] uses the above formula to compute the power of the test for the difference between two means.

Example 5.9.3 (Power Calculation) Suppose a researcher hypothesizes that the mean $\lambda_1 = 3$ of a Poisson population is at least one unit larger than the mean λ_2 of another population, and he likes to test

$$H_0 : \lambda_1 - \lambda_2 \le 0 \quad \text{vs.} \quad H_a : \lambda_1 - \lambda_2 > 0.$$

To find the required sample size to get a power of 0.80 at level of 0.05, enter 30 for both sample sizes, 0 for d in [H0: M1-M2 = d], 1 for one-tail test, 0.05 for level, 3 for [Guess M1], 2 for [Guess M2] and click [Power] to get 0.791813. By raising the sample size to 31, we get a power of 0.803148. We also note that when the first sample size is 27 and the second sample size is 36, the power is 0.803128.

For the above example, if it is desired to find the sample sizes for testing the hypotheses

$$H_0 : \lambda_1 - \lambda_2 = 0 \quad \text{vs.} \quad H_a : \lambda_1 - \lambda_2 \ne 0,$$

then enter 2 for two-tail test (while keep the other values as they are), and click [Power]. For example, when both sample sizes are 33, the power is 0.730551; when they are 39, the power is 0.800053. [Note that if one choose to use the conditional test, then the required sample size from both populations is 41. See Example 5.7.3].

5.10 Model Fitting with Examples

Example 5.10.1 Rutherford and Geiger (1910) presented data on α particles emitted by a radioactive substance in 2608 periods, each of 7.5 sec. The data are given in Table 5.1.

a. Fit a Poisson model for the data.

b. Estimate the probability of observing 5 or less α particles in a period of 7.5 sec.

c. Find a 95% confidence interval for the mean number of α particles emitted in a period of 7.5 sec.

Table 5.1 Observed frequency O_x of the number of α particles x in 7.5 second periods

x	0	1	2	3	4	5	6	7	8	9	10
O_x	57	203	383	525	532	408	273	139	45	27	16
E_x	54.6	211	408	526	508	393	253	140	67.7	29.1	17

Solution:

a. To fit a Poisson model, we estimate first the mean number λ of α particles emitted per 7.5 second period. Note that

$$\widehat{\lambda} = \frac{1}{2608}\sum_{x=0}^{10} xO_x = \frac{10086}{2608} = 3.867.$$

Using this estimated mean, we can compute the probabilities and the expected (theoretical) frequencies E_x under the Poisson($\widehat{\lambda}$) model. For example, E_0 is given by

$$E_0 = P(X = 0|\lambda = 3.867) \times 2608 = 0.020921 \times 2608 = 54.6.$$

Other expected frequencies can be computed similarly. These expected frequencies are given in Table 5.1. We note that the observed and the expected frequencies are in good agreement. Furthermore, for this example, the chi-square statistic

$$\chi^2 = \sum_{x=0}^{10} \frac{(O_x - E_x)^2}{E_x} = 13.06,$$

and the df $= 11 - 1 - 1 = 9$ (see Section 1.4.2). The p-value for testing

$$H_0\text{: The data fit Poisson(3.867) model vs. } H_a\text{: } H_0 \text{ is not true}$$

is given by $P(\chi_9^2 > 13.06) = 0.16$, which implies that the Poisson(3.867) model is tenable for the data.

b. Select the dialog box [StatCalc→Discrete→Poisson→Probabilities, Critical Values and Moments], enter 3.867 for the mean, and 5 for k; click [P(X <= k)] to get

$$P(X \leq 5) = \sum_{k=0}^{5} \frac{e^{-3.867}(3.867)^k}{k!} = 0.805557.$$

c. To compute the 95% confidence interval for the mean, select the dialog box [StatCalc→Discrete→Poisson→CI for Mean and Sample Size for Width] from *StatCalc*, enter 10086 for the observed k, 2608 for the sample size n and 0.95 for the confidence level; click [2-sided] to get (3.79222, 3.94356).

Example 5.10.2 Data on the number of deaths due to kicks from horses, based on the observation of 10 Prussian cavalry corps for 20 years (equivalently, 200 corps-years), are given in Table 5.2. Prussian officials collected this data during the latter part of the 19th century in order to study the hazards that horses posed to soldiers (Bortkiewicz 1898).

Table 5.2 Horse kick data

Number of deaths k:	0	1	2	3	4	5
Number of corps-years in which k deaths occurred, O_x:	109	65	22	3	1	0
Expected Number of corps-years, E_x:	108.7	66.3	20.2	4.1	0.6	0

In this situation, the chances of death due to a kick from horse is small while the number of soldiers exposed to the risk is quite large. Therefore, a Poisson distribution may well fit the data. As in Example 5.10.1, the mean number of deaths per period can be estimated as

$$\widehat{\lambda} = \frac{0 \times 109 + 1 \times 65 + \ldots + 5 \times 0}{200} = 0.61.$$

Using this estimated mean, we can compute the expected frequencies as in Example 5.10.1. They are given in the third row of Table 5.2. For example, the expected frequency in the second column can be obtained as

$$P(X = 1|\lambda = 0.61) \times 200 = 0.331444 \times 200 = 66.3.$$

We note that the observed and the expected frequencies are in good agreement. Furthermore, for this example, the chi-square statistic

$$\chi^2 = \sum_{x=0}^{5} \frac{(O_x - E_x)^2}{E_x} = 0.7485,$$

and the df = 4. The p-value for testing

$$H_0\text{: The data fit Poisson(0.61) model vs. } H_a\text{: } H_0 \text{ is not true}$$

is given by $P(\chi^2_4 > 0.7485) = 0.9452$ which is greater than any practical level of significance. Therefore, the Poisson(0.61) model is tenable. A 95% confidence interval for the mean number of deaths is (0.51, 0.73). To get the confidence interval, select [StatCalc→Discrete→Poisson→CI for Mean and Sample Size for Width] from *StatCalc*, enter 200 for the sample size, 122 for total count, 0.95 for the confidence level; click [2-sided].

5.11 Properties and Results

5.11.1 Properties

1. For a fixed k, $P(X \le k|\lambda)$ is a nonincreasing function of λ.

2. Let $X_1, \ldots, X_n$ be independent Poisson random variables with $E(X_i) = \lambda_i$, $i = 1, \ldots, n$. Then
$$\sum_{i=1}^{n} X_i \sim \text{Poisson}\left(\sum_{i=1}^{n} \lambda_i\right).$$

3. Recurrence Relations:
$$\begin{aligned} P(X = k+1|\lambda) &= \tfrac{\lambda}{k+1} P(X = k|\lambda), \quad k = 0, 1, 2, \ldots \\ P(X = k-1|\lambda) &= \tfrac{k}{\lambda} P(X = k|\lambda), \quad k = 1, 2, \ldots \end{aligned}$$

4. An identity: Let X be a Poisson random variable with mean λ and $|g(-1)| < \infty$. Then,
$$E[Xg(X-1)] = \lambda E[g(X)]$$
provided the indicated expectations exist. [Hwang 1982]

5.11.2 Relation to Other Distributions

1. Binomial: Let X_1 and X_2 be independent Poisson random variables with means λ_1 and λ_2 respectively. Then, conditionally
$$X_1|(X_1 + X_2 = n) \sim \text{binomial}\left(n, \frac{\lambda_1}{\lambda_1 + \lambda_2}\right).$$

2. Multinomial: If $X_1, \ldots, X_m$ are independent Poisson(λ) random variables, then the conditional distribution of X_1 given $X_1 + \ldots + X_m = n$ is multinomial with n trials and cell probabilities $p_1 = \ldots = p_m = 1/m$.

3. Gamma: Let X be a Poisson(λ) random variable. Then

$$P(X \leq k|\lambda) = P(Y \geq \lambda),$$

where Y is Gamma($k+1, 1$) random variable. Furthermore, if W is a gamma(a, b) random variable, where a is an integer, then for $x > 0$,

$$P(W \leq x) = P\left(Q \geq a\right),$$

where Q is a Poisson(x/b) random variable.

5.11.3 Approximations

1. Normal:

$$\begin{aligned} P(X \leq k|\lambda) &\simeq P\left(Z \leq \tfrac{k-\lambda+0.5}{\sqrt{\lambda}}\right), \\ P(X \geq k\lambda) &\simeq P\left(Z \geq \tfrac{k-\lambda-0.5}{\sqrt{\lambda}}\right), \end{aligned}$$

where X is the Poisson(λ) random variable and Z is the standard normal random variable.

5.12 Random Number Generation

```
Input:
        L = Poisson mean
        ns = desired number of random numbers

Output:
        x(1),..., x(ns) are random numbers from the
        Poisson(L) distribution
```

The following algorithm is based on the inverse method, and is similar to Algorithm 3.9.1 for the binomial random numbers generator.

Algorithm 5.12.1

```
Set k = int(L); pk = P(X = k); df = P(X <= k)
    rpk = pk; rk = k
    max = L + 10*sqrt(L)
    If L > 100,  max = L + 6*sqrt(L)
    If L > 1000, max = L + 5*sqrt(L)

For j = 1 to ns
      Generate u from uniform(0, 1)
      If u > df, go to 2
1     u = u + pk
      If k = 0 or u > df, go to 3
      pk = pk*k/L
      k = k - 1
      go to 1
2     pk = L*pk/(k + 1)
      u = u - pk
      k = k + 1
      If k = max or u < df, go to 3
      go to 2
3     x(j) = k
      k = rk
      pk = rpk
[end j loop]
```

5.13 Computation of Probabilities

For a given k and small mean λ, $P(X = k)$ can be computed in a straightforward manner. For large values, the logarithmic gamma function can be used.

To Compute $P(X = k)$:

$$P(X = k) = \exp(-\lambda + k * \ln(\lambda) - \ln(\Gamma(k + 1))$$

To compute $P(X \leq k)$:

Compute $P(X = k)$
Set $m = \text{int}(\lambda)$

If $k \le m$, compute the probabilities using the backward recursion relation

$$P(X = k-1|\lambda) = \frac{k}{\lambda} P(X = k|\lambda),$$

for $k-1, k-2, \ldots, 0$ or until the desired accuracy; add these probabilities and $P(X = k)$ to get $P(X \le k)$.
else compute the probabilities using the forward recursion relation

$$P(X = k+1|\lambda) = \frac{\lambda}{k+1} P(X = k|\lambda),$$

for $k+1, k+2, \ldots$ until the desired accuracy; sum these probabilities to get $P(X \ge k+1)$; the cumulative probability $P(X \le k) = 1 - P(X \ge k+1)$.

The following algorithm for computing a Poisson cdf is based on the above method.

Algorithm 5.13.1

```
Input:
       k = the nonnegative integer at which the cdf is to be evaluated
      el = the mean of the Poisson distribution, el > 0
Output:
       poicdf = P(X <= k|el)

Set  mode = int(el)
     one = 1.0d0

     if(k .lt. 0) return poicdf = 0.0d0

     pk = poiprob(k, el)
     poicdf = 0.0d0
     i = k
     if(k .le. mode) then
1       poicdf = poicdf + pk;
        pk = pk*i/el
        if(i .eq. 0 .or. pk .lt. 1.0d-14) return
        i = i - 1
        goto 1
     else
2       pk = pk*el/(i+one)
        if(pk. lt. 1.0d-14) goto 3
```

```
         poicdf = poicdf + pk
         i = i + 1
         goto 2
3        poicdf = one-poicdf
      end if
```

The following Fortran routines compute the cdf (poicdf) and pmf (poiprob) of a Poisson distribution with mean λ = el.

```
Input:
      k = the value at which the cdf is to be evaluated
     el = mean of the Poisson distribution

Output:
      P(X <= x) = poicdf(k, el)

cccccccccccccccccccccccccccccccccccccccccccccccccccc
      double precision function poicdf(k, el)
      implicit doubleprecision (a-h,o-z)
      data zero, one/0.0d0, 1.0d0/

      poicdf  = zero
      if(k .lt. 0) return

      mode = int(el)

      pk = poiprob(k, el)
      i = k
      if(k .le. mode) then
1        poicdf = poicdf + pk;
         pk = pk*i/el
         if(i .eq. 0 .or. pk .lt. 1.0d-14) return
         i = i - 1
         goto 1
      else
2        pk = pk*el/(i+one)
         if(pk. lt. 1.0d-14) goto 3
         poicdf = poicdf + pk
         i = i + 1
         goto 2
```

```
3        poicdf = one - poicdf
      end if
      end


cccccccccccccccccccccccccccccccccccccccccccccccccccc
      double precision function poiprob(k, el)
      implicit doubleprecision (a-h,o-z)
      data zero, one/0.0d0, 1.0d0/

      poiprob = zero
      if(k .lt. 0) return
c
c alng(x) = logarithmic gamma function given in Section 1.8
c
      term1 = -alng(k+one)+k*dlog(el)-el
      poiprob = dexp(term1)
      end
```

Chapter 6

Geometric Distribution

6.1 Description

Consider a sequence of independent Bernoulli trials with success probability p. Let X denote the number of failures until the first success to occur. Then, the probability mass function of X is given by

$$\begin{aligned} P(X = k|p) &= P(\text{Observing } k \text{ failures}) \\ &\times P(\text{Observing a success at } (k+1)\text{st trial}) \\ &= (1-p)^k p, \quad k = 0, 1, 2, \ldots \end{aligned}$$

This is the probability of observing exactly k failures until the first success to occur or the probability that exactly $(k+1)$ trials are required to get the first success. The cdf is given by

$$F(k|p) = p \sum_{i=0}^{k} (1-p)^i = \frac{p[1-(1-p)^{k+1}]}{1-(1-p)} = 1-(1-p)^{k+1}, \quad k = 0, 1, 2, \ldots$$

Since the above cdf is a geometric series with finite terms, the distribution is called geometric distribution.

6.2 Moments

Mean:	$\frac{1-p}{p}$
Variance:	$\frac{1-p}{p^2}$
Mode:	0
Mean Deviation:	$2u(1-p)^u$, where u is the smallest integer greater than the mean.
Coefficient of Variation:	$\frac{1}{\sqrt{(1-p)}}$
Coefficient of Skewness:	$\frac{2-p}{\sqrt{(1-p)}}$
Coefficient of Kurtosis:	$9 + \frac{p^2}{(1-p)}$
Moments about the Mean:	$\mu_{k+1} = (1-p)\left(\frac{\partial \mu_k}{\partial q} + \frac{k}{p^2}\mu_{k-1}\right)$, where $q = 1-p$, $\mu_0 = 1$ and $\mu_1 = 0$.
Moment Generating Function:	$p(1-qe^t)^{-1}$
Probability Generating Function:	$p(1-qt)^{-1}$

6.3 Computing Table Values

The dialog box [StatCalc→Discrete→Geometric] computes the tail probabilities, critical points, parameters and confidence intervals for a geometric distribution with parameter p.

To compute probabilities: Enter the number of failures k until the first success and the success probability p; click [P].

Example 6.3.1 The probability of observing the first success at the 12th trial, when the success probability is 0.1, can be computed as follows: Enter 11 for k, and 0.1 for p; click on [P] to get

$$P(X \leq 11) = 0.71757, \ P(X \geq 11) = 0.313811 \text{ and } P(X = 11) = 0.0313811.$$

Example 6.3.2 To find the value of the success probability when $k = 11$ and P(X <= k) = 0.9, enter these numbers in the appropriate edit boxes, and click [s] to get 0.174596.

Example 6.3.3 To find the value of k when $p = 0.3$ and P(X <= k) = 0.8, enter these numbers in the white boxes, and click [k] to get 4.

To compute confidence intervals: Enter the observed number of failures k until the first success and the confidence level; click on [1-sided] to get one-sided limits or click [2-sided] to get two-sided confidence intervals.

Example 6.3.4 Suppose that in an experiment consisting of sequence of Bernoulli trials, 12 trials were required to get the first success. To find a 95% confidence interval for the success probability p, enter 11 for k, 0.95 for confidence level; click [2-sided] to get (0.002, 0.285).

To compute moments: Enter a value for p in $(0, 1)$; click [M].

6.4 Properties and Results

1. $P(X \geq k+1) = (1-p)^{k+1}, \quad k = 0, 1, \ldots$

2. For fixed k, $P(X \leq k|p)$ is an increasing function of p.

3. *Memoryless Property:* For nonnegative integers k and m,

$$P(X > m + k | X > m) = P(X > k).$$

   The probability of observing an additional k failures, given the fact that m failures have already observed, is the same as the probability of observing k failures at the start of the sequence. That is, geometric distribution forgets what has occurred earlier.

4. If $X_1, \ldots, X_r$ are independent geometric random variables with success probability p, then

$$\sum_{i=1}^{r} X_i \sim \text{negative binomial}(r, p).$$

6.5 Random Number Generation

Generate u from uniform(0, 1)
Set k = integer part of $\frac{\ln(u)}{\ln(1-p)}$
k is a pseudo random number from the geometric(p) distribution.

Chapter 7

Negative Binomial Distribution

7.1 Description

Consider a sequence of independent Bernoulli trials with success probability p. The distribution of the random variable that represents the number of failures until the first success is called geometric distribution. Now, let X denote the number of failures until the rth success. The random variable X is called the negative binomial random variable with parameters p and r, and its pmf is given by

$$\begin{aligned} P(X = k|r, p) &= P(\text{observing } k \text{ failures in the first } k + r - 1 \text{ trials}) \\ &\times\ P(\text{observing a success at the } (k + r)\text{th trial}) \\ &= \binom{r+k-1}{k} p^{r-1}(1-p)^k \times p. \end{aligned}$$

Thus,

$$f(k|r, p) = P(X = k|r, p) = \binom{r+k-1}{k} p^r(1-p)^k, \quad k = 0, 1, 2, \ldots; \quad 0 < p < 1.$$

This is the probability of observing k failures before the rth success or equivalently, probability that $k + r$ trials are required until the rth success to occur. In the binomial distribution, the number of successes out of fixed number of trials is a random variable whereas in the negative binomial the number of trials required to have a given number of successes is a random variable. The following relation between the negative binomial and binomial distributions is worth noting.

$$\begin{aligned}
P(X \le k) &= P(\text{observing } k \text{ or less failures before the } r\text{th success}) \\
&= P((k+r) \text{ or less trials are required to have exactly } r \text{ successes}) \\
&= P(\text{observing } r \text{ or more successes in } (k+r) \text{ trials}) \\
&= \sum_{i=r}^{k+r} \binom{k+r}{i} p^i (1-p)^{k+r-i} \\
&= P(Y \ge r),
\end{aligned}$$

where Y is a binomial$(k+r, p)$ random variable.

The plots of the probability mass functions presented in Figure 7.1 show that the negative binomial distribution is always skewed to the right. The degree of skewness decreases as r increases. See the formula for coefficient of skewness in Section 7.2.

7.2 Moments

Mean:	$\frac{r(1-p)}{p}$
Variance:	$\frac{r(1-p)}{p^2}$
Mode:	The largest integer $\le \frac{(r-1)(1-p)}{p}$.
Mean Deviation:	$2u\binom{r+u-1}{u}(1-p)^u p^{r-1}$, where u is the smallest integer greater than the mean. [Kamat 1965]
Coefficient of Variation:	$\frac{1}{\sqrt{r(1-p)}}$
Coefficient of Skewness:	$\frac{2-p}{\sqrt{r(1-p)}}$
Coefficient of Kurtosis:	$3 + \frac{6}{r} + \frac{p^2}{r(1-p)}$
Central Moments:	$\mu_{k+1} = q\left(\frac{\partial \mu_k}{\partial q} + \frac{kr}{p^2}\mu_{k-1}\right)$, where $q = 1-p$, $\mu_0 = 1$ and $\mu_1 = 0$.

Moment Generating Function:	$p^r(1-qe^t)^{-r}$
Probability Generating Function:	$p^r(1-qt)^{-r}$

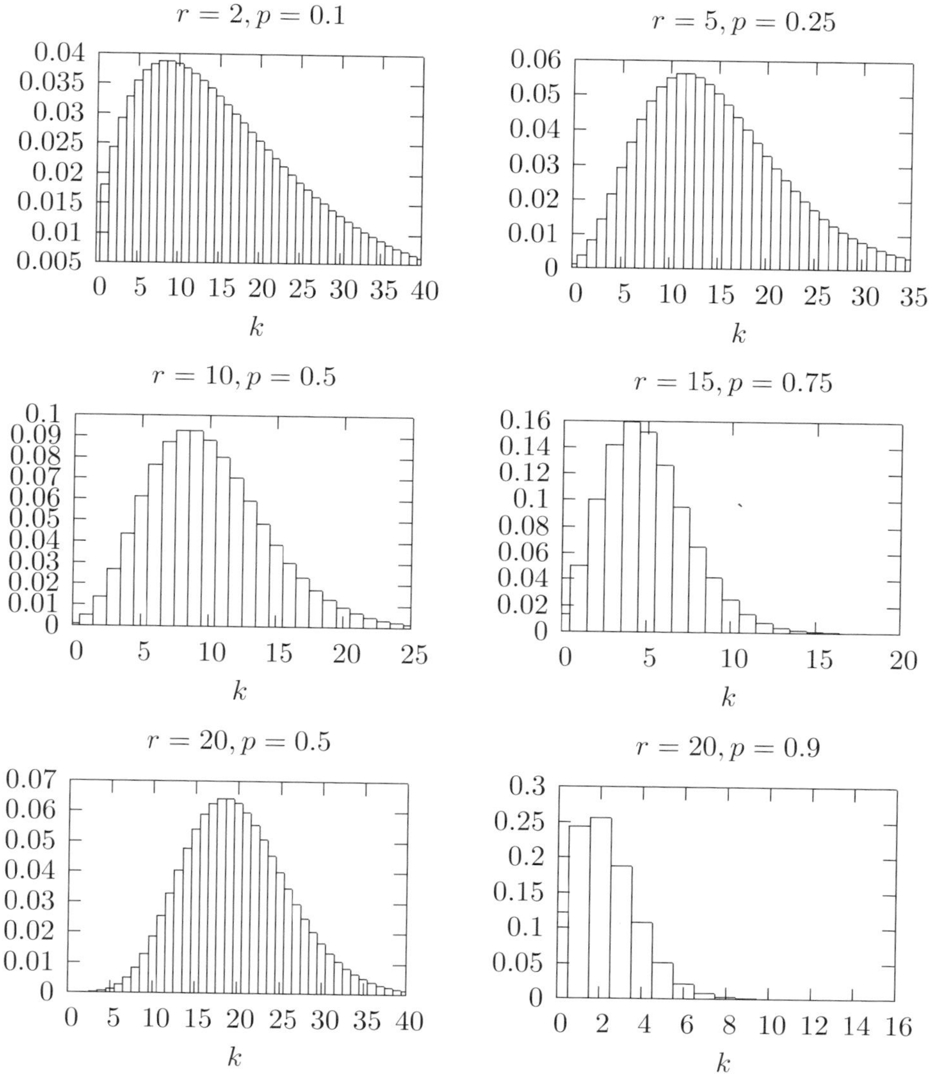

Figure 7.1 Negative Binomial Probability Mass Functions; k is the Number of Failures until the rth Success

7.3 Computing Table Values

The dialog box [StatCalc→Discrete→Negative Binomial] computes the following table values.

1. Tail Probabilities, Critical Values and Moments.
2. Test for Proportion [Section 7.5].
3. Confidence Interval for Proportion [Section 7.6].

To compute probabilities: Enter the number of successes r, number of failures until the rth success, and the success probability; click [P].

Example 7.3.1 When $r = 20, k = 18$, and $p = 0.6$,

$$P(X \leq 18) = 0.862419, \; P(X \geq 18) = 0.181983, \text{ and } P(X = 18) = 0.0444024.$$

Example 7.3.2 To find the success probability when $k = 5$, P(X <= k) = 0.56, and $r = 4$, enter these values in appropriate edit boxes, and click [s] to get 0.417137.

To compute moments: Enter the values of r and the success probability p; click [M].

Illustrative Examples

Example 7.3.3 A coin is to be flipped sequentially.

a. What are the chances that the 10th head will occur at the 12th flip?

b. Suppose that the 10th head had indeed occurred at the 12th flip. What can be said about the coin?

Solution:

a. Let us assume that the coin is balanced. To find the probability, select the dialog box [StatCalc→Discrete→Negative Binomial→Probabilities, Critical Values and Moments] from *StatCalc*, enter 10 for the number of successes, 2 for the number of failures, and 0.5 for the success probability; click [$P(X <= k)$] to get 0.01343.

b. If the coin were balanced, then the probability of observing 2 or less tails before the 10th head is only 0.01929, which is less than 2%. Therefore, if one observes 10th head at the 12th flip, then it indicates that the coin is not balanced. To find this probability using *StatCalc*, just follow the steps of part (a).

Example 7.3.4 A shipment of items is submitted for inspection. In order to save the cost of inspection and time, the buyer of the shipment decided to adopt the following acceptance sampling plan: He decided to inspect a sample of not more than thirty items. Once the third defective observed, he will stop the inspection and reject the lot; otherwise he will continue the inspection up to the 30th item. What are the chances of rejecting the lot if it indeed contains 15% defective items?

Solution: Let X denote the number of nondefective items that must be examined in order to get 3 or more defective items. If we refer defective as success and nondefective as failure, then X follows a negative binomial with $p = 0.15$ and $r = 3$. We need to find the probability of observing 27 or less nondefective items to get the third defective item. Thus, the required probability is

$$P(X \leq 27|3,\ 0.15) = 0.8486,$$

which can be computed using *StatCalc* as follows: Select the dialog box [StatCalc→Discrete→Negative Binomial→Probabilities, Critical Values and Moments] from *StatCalc*, enter 3 for the number of successes, 27 for the number of failures, and 0.15 for the success probability; click [P(X <= k)] to get 0.848599. Thus, for this acceptance sampling plan, the chances of rejecting the lot is about 85% if the lot actually contains 15% defective items.

7.4 Point Estimation

Suppose that a binomial experiment required $k + r$ trials to get the rth success. Then the uniformly minimum variance unbiased estimator of the success probability is given by

$$\hat{p} = \frac{r-1}{r+k-1},$$

and its approximate variance is given by

$$\mathrm{Var}(\hat{p}) \simeq \frac{p^2(1-p)}{2}\left(\frac{2k+2-p}{k(k-p+2)}\right).$$

7.5 A Test for the Proportion

Suppose that in a sequence of independent Bernoulli trials, each with success probability p, rth success was observed at the $(k + r)$th trial. Based on this information, we like to test about the true success probability p.

For testing

$$H_0 : p \le p_0 \quad \text{vs.} \quad H_a : p > p_0, \tag{7.5.1}$$

the null hypothesis will be rejected if the p-value $P(X \le k|r, p_0) \le \alpha$, for testing

$$H_0 : p \ge p_0 \quad \text{vs.} \quad H_a : p < p_0, \tag{7.5.2}$$

the null hypothesis will be rejected if the p-value $P(X \ge k|r, p_0) \le \alpha$ and for testing

$$H_0 : p = p_0 \quad \text{vs.} \quad H_a : p \ne p_0, \tag{7.5.3}$$

the null hypothesis will be rejected if the p-value

$$2 \min\{P(X \le k|r, p_0), P(X \ge k|r, p_0)\} \le \alpha.$$

The dialog box [StatCalc→Discrete→Negative Binomial→Test for p] computes the above p-values for testing the success probability.

Example 7.5.1 A shipment of items is submitted for inspection. The buyer of the shipment inspected the items one-by-one randomly, and found the 5th defective item at the 30th inspection. Based on this information, can we conclude that the proportion of defective items p in the shipment is less than 30%? Find a point estimate of p.

Solution: Let p denote the true proportion of defective items in the shipment. Then, we want to test

$$H_0 : p \ge 0.3 \quad \text{vs.} \quad H_a : p < 0.3.$$

To compute the p-value, select the dialog box [StatCalc→Discrete→ Negative Binomial→Test for p], enter 5 for r, 25 for k, 0.3 for [Value of p0] and click [p-values for] to get 0.0378949. This is the p-value for the left-tail test, and is less than 0.05. Therefore, we conclude that the true proportion of defective items in the shipment is less than 30%. A point estimate of the actual proportion of defective items is

$$\hat{p} = \frac{r-1}{r+k-1} = \frac{5-1}{5+25-1} = 0.1379.$$

Suppose one inadvertently applies the binomial testing method described in Section 3.4 instead of negative binomial, then $n = 30$, and the number of successes is 5. Using [StatCalc→Discrete→Binomial→Test for p and Power Calculation], we get the p-value for testing above hypotheses as 0.0765948. Thus, on contrary to the result based on the negative binomial, the result based on the binomial is not significant at 0.05 level.

7.6 Confidence Intervals for the Proportion

For a given r and k, an exact $1-\alpha$ confidence interval for p can be computed using Clopper–Pearson approach. The lower limit p_L satisfies

$$P(X \le k|r, p_L) = \alpha/2,$$

and the upper limit p_U satisfies

$$P(X \ge k|r, p_U) = \alpha/2.$$

Using the relation between negative binomial and beta random variables (see Section 16.6.2), it can be shown that

$$P_L = \text{beta}^{-1}(\alpha/2; r, k+1)$$

and

$$p_U = \text{beta}^{-1}(1-\alpha/2; r, k),$$

where $\text{beta}^{-1}(c; a, b)$ denotes the cth quantile of the beta distribution with shape parameters a and b.

The dialog box [StatCalc→Discrete→Negative Binomial→CI for p and Sample Size for Precision] uses the above methods to compute confidence intervals for p.

Example 7.6.1 A shipment of items is submitted for inspection. The buyer of the shipment inspected the items one-by-one randomly, and found the 6th defective item at the 30th inspection. Based on this information, find a 95% confidence interval for the true proportion of defective items in the shipment.

Solution: To find the 95% exact confidence interval for the true proportion of defective items, enter 6 for r, 24 for k, 0.95 for confidence level, and click [2-sided] to get (0.0771, 0.3577). That is, the true percentage of defective items in the shipment is between 7.7 and 36 with confidence 95%.

7.7 Properties and Results

In the following X denotes the negative binomial(r, p) random variable.

7.7.1 Properties

1. For a given k and r, $P(X \le k)$ is a nondecreasing function of p.

2. Let $X_1, \ldots, X_m$ be independent negative binomial random variables with

$$X_i \sim \text{negative binomial}(r_i, p), \quad i = 1, 2, ..., m.$$

Then,

$$\sum_{i=1}^{m} X_i \sim \text{negative binomial}\left(\sum_{i=1}^{m} r_i, p\right).$$

3. Recurrence Relations:

$$\begin{aligned} P(X = k+1) &= \tfrac{(r+k)(1-p)}{(k+1)} P(X = k) \\ P(X = k-1) &= \tfrac{k}{(r+k-1)(1-p)} P(X = k) \end{aligned}$$

7.7.2 Relation to Other Distributions

1. Binomial: Let X be a negative binomial(r, p) random variable. Then

$$P(X \le k | r, p) = P(Y \ge r), \ k = 1, 2, ...$$

where Y is a binomial random variable with $k + r$ trials and success probability p.

2. Beta: See Section 16.6.2.

3. Geometric Distribution: Negative binomial distribution with $r = 1$ specializes to the geometric distribution described in Chapter 6.

7.8 Random Number Generation

```
Input:
      r = number of successes; p = success probability
      ns = desired number of random numbers

Output:
      k = random number from the negative binomial(r, p)
      distribution; the number of failures until the rth
      success
```

Algorithm 7.8.1

```
Set i = 1; k = 0

For j = 1 to ns
1       Generate u from uniform(0, 1)
        If u <= p, k = k + 1
        If k = r goto 2
        i = i + 1
        go to 1
2       x(j) = i - r
        k = 0
        i = 1
[end j loop]
```

The following algorithm is based on the inverse method, and is similar to Algorithm 3.9.1 for binomial variates generator.

Algorithm 7.8.2

```
Set     k = int((r - 1.0)*(1 - p)/p)
        pk = P(X = k|r, p)
        df = P(X <= k|r, p)
        rpk = pk
        ik = k
        xb = r*(1 - p)/p
        s  = sqrt(xb/p)
        mu = xb + 10.0*s
        if(xb > 30.0) mu = xb + 6.0*s
        if(xb > 100.0) mu = xb + 5.0*s
        ml = max(0.0, mu - 10.0*s)
        if(xb > 30.0)  ml = max(0.0, xb - 6.0*s)
        if(xb > 100.0) ml = max(0.0, xb - 5.0*s)
For j = 1 to ns
        Generate u from uniform(0, 1)
        if(u > df) goto 2
1       u = u + pk
        if(k = ml or u > df) goto 3
        pk = pk*k/((r + k - 1.0)*(1.0 - p))
        k = k - 1
```

```
        goto 1
2       pk = (r + k)*(1 - p)*pk/(k + 1)
        u = u - pk
        k = k + 1
        If k = mu or u <= df, go to 3
        go to 2
3       x(j) = k
        k = rk
        pk = rpk
[end j loop]
```

7.9 A Computational Method for Probabilities

For small values of k and r, $P(X = k)$ can be computed in a straightforward manner. For other values, logarithmic gamma function $\ln\Gamma(x)$ given in Section 1.8 can be used.

To compute $P(X = k)$:

Set $q = 1 - p$
$c = \ln\Gamma(r + k) - \ln\Gamma(k + 1) - \ln\Gamma(r)$
$b = k\ln(q) + r\ln(p)$
$P(X = k) = \exp(c + b)$

To compute $P(X \leq k)$:

If an efficient algorithm for evaluating the cdf of beta distribution is available, then the following relation between the beta and negative binomial distributions,

$$P(X \leq k) = P(Y \leq p),$$

where Y is a beta variable with shape parameters r and $k + 1$, can be used to compute the cumulative probabilities.

The relation between the binomial and negative binomial distributions,

$$P(X \leq k) = 1.0 - P(W \leq r - 1),$$

where W is a binomial random variable with $k+r$ trials and success probability p, can also be used to compute the cumulative probabilities.

Chapter 8

Logarithmic Series Distribution

8.1 Description

The probability mass function of a logarithmic series distribution with parameter θ is given by

$$P(X = k) = \frac{a\theta^k}{k}, \quad 0 < \theta < 1, \ k = 1, 2, \ldots,$$

where $a = -1/[\ln(1-\theta)]$; the cdf is given by

$$F(k|\theta) = P(X \le k|\theta) = a \sum_{i=1}^{k} \frac{\theta^k}{k}, \quad 0 < \theta < 1, \ k = 1, 2, \ldots$$

The logarithmic series distribution is useful to describe a variety of biological and ecological data. Specifically, the number of individuals per species can be modeled using a logarithmic series distribution. This distribution can also be used to fit the number of products requested per order from a retailer. Williamson and Bretherton (1964) used a logarithmic series distribution to fit the data that represent quantities of steel per order from a steel merchant; they also tabulated the cumulative probabilities for various values of the mean of the distribution. Furthermore, Chatfield et al. (1966) fitted the logarithmic series distribution to the distribution of purchases from a random sample of consumers.

The logarithmic series distribution is always right-skewed (see Figure 8.1).

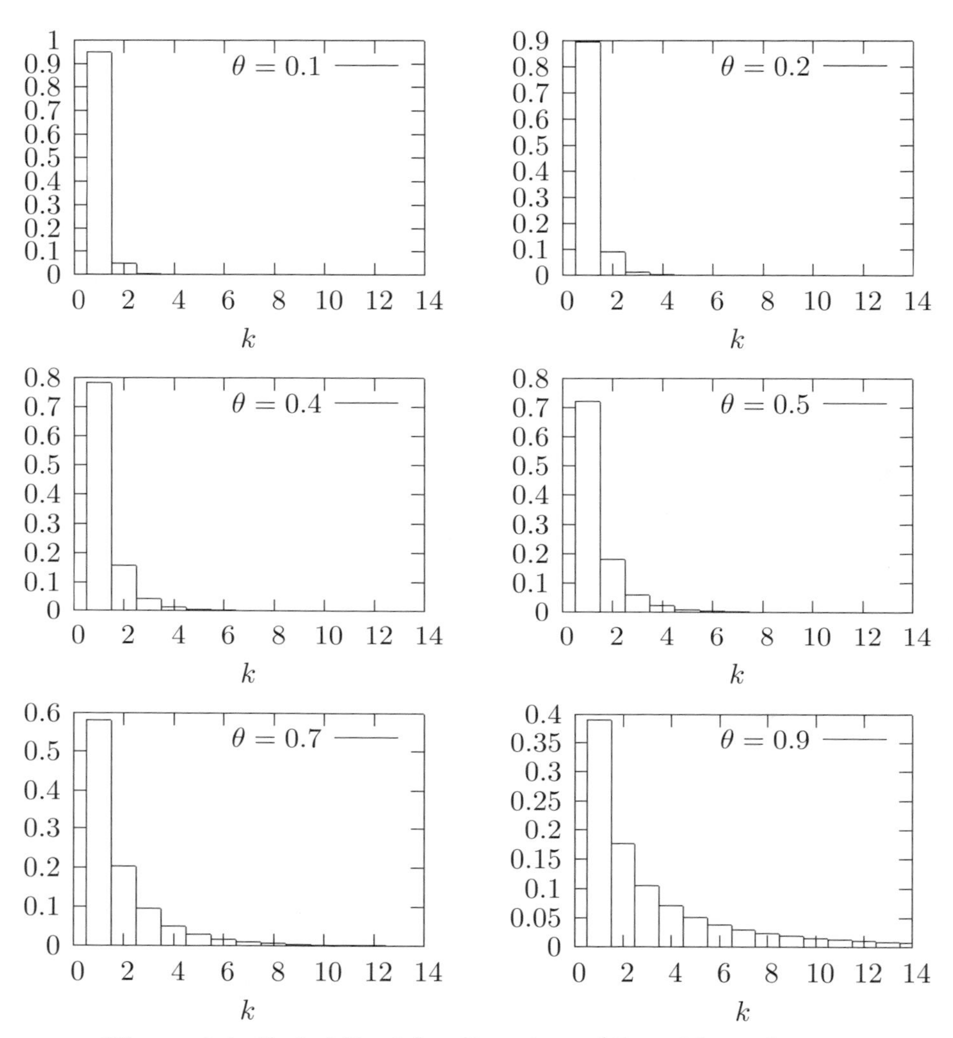

Figure 8.1 Probability Mass Functions of Logarithmic Series

8.2 Moments

Let $a = -1/[\ln(1-\theta)]$.

Mean:	$\frac{a\theta}{1-\theta}$
Variance:	$\frac{a\theta(1-a\theta)}{(1-\theta)^2}$
Coefficient of Variation:	$\sqrt{\frac{(1-a\theta)}{a\theta}}$
Coefficient of Skewness:	$\frac{a\theta(1+\theta-3a\theta+2a^2\theta^2)}{[a\theta(1-a\theta)]^{3/2}}$
Coefficient of Kurtosis:	$\frac{1+4\theta+\theta^2-4a\theta(1+\theta)+6a^2\theta^2-3a^3\theta^{\,3}}{a\theta(1-a\theta)^2}$
Mean Deviation:	$\frac{2a\theta(\theta^m-P(X>m))}{1-\theta}$, where m denotes the largest integer $\leq$ the mean. [Kamat 1965]
Factorial Moments:	$E\left(\prod_{i=1}^{k}(X-i+1)\right) = \frac{a\theta^k(k-1)!}{(1-\theta)^k}$
Moment Generating Function:	$\frac{\ln(1-\theta\exp(t))}{\ln(1-\theta)}$
Probability Generating Function:	$\frac{\ln(1-\theta t)}{\ln(1-\theta)}$

8.3 Computing Table Values

The dialog box [StatCalc→Discrete→Logarithmic Series] in *StatCalc* computes the probabilities, moments, and the maximum likelihood estimator of θ based on a given sample mean.

To compute probabilities: Enter the values of the parameter θ and the observed value k; click [P(X <= k)].

Example 8.3.1 When $\theta = 0.3$ and $k = 3$,

$$P(X \leq 3) = 0.9925, \quad P(X \geq 3) = 0.032733 \ \text{ and } \ P(X = 3) = 0.025233.$$

StatCalc also computes the value of θ or the value of k. For example, when $k = 3$, $P(X \leq 3) = 0.6$, the value of θ is 0.935704. To get this value, enter 3 for k, 0.6 for [P(X <= k)] and click [T].

To compute the MLE of θ: Enter the sample mean, and click on [MLE].

Example 8.3.2 When the sample mean = 2, the MLE of θ is 0.715332.

To compute moments: Enter a value for θ in (0,1); click [M].

Example 8.3.3 A mail-order company recorded the number of items purchased per phone call or mail in form. The data are given in Table 8.1. We will fit a logarithmic series distribution for the number of item per order. To fit the model, we first need to estimate the parameter θ based on the sample mean which is

$$\bar{x} = \frac{\sum x_i f_i}{\sum f_i} = \frac{2000}{824} = 2.4272.$$

To find the MLE of θ using *StatCalc*, enter 2.4272 for the sample mean, and click [MLE] to get 0.7923. Using this number for the value of θ, we can compute the probabilities $P(X = 1)$, $P(X = 2)$, etc. These probabilities are given in column 3 of Table 8.1. To find the expected frequencies, multiply the probability by the total frequency, which is 824 for this example.

Comparison of the observed and expected frequencies indicates that the logarithmic series distribution is very well fitted for the data. The fitted distribution can be used to check whether the distribution of number of items demanded per order changes after a period of time.

Example 8.3.4 Suppose that the mail-order company in the previous example collected new data after a few months after the previous study, and recorded them as shown Table 8.2.

First we need to check whether a logarithmic series distribution still fits the data. The sample mean is

$$\bar{x} = \frac{\sum x_i f_i}{\sum f_i} = \frac{1596}{930} = 1.7161,$$

and using *StatCalc*, we find that the MLE of θ is 0.631316. As in Example 8.3.3, we can compute the probabilities and the corresponding expected frequencies using 0.631316 as the value of θ. Comparison of the observed frequencies with the expected frequencies indicate that a logarithmic series distribution still fits the data well; however, the smaller MLE indicates that the demand for fewer units per order has increased since the last study.

Table 8.1 Number of items ordered per call or mail-in form

No. of items x_i	Observed frequency	Probability	Expected frequency
1	417	0.504116	415.4
2	161	0.199706	164.6
3	84	0.105485	86.9
4	50	0.062682	51.6
5	39	0.039730	32.7
6	22	0.026232	21.6
7	12	0.017814	14.7
8	8	0.012350	10.2
9	7	0.008698	7.2
10	6	0.006202	5.1
11	5	0.004467	3.7
12 and over	13	0.012518	10.3
Total	824	1.0	824

Table 8.2 Number of items ordered per call or mail-in form after a few months

No. of items x_i	Observed frequency	Probability	Expected frequency
1	599	0.632698	588.4
2	180	0.199716	185.7
3	75	0.084056	78.2
4	30	0.039799	37
5	20	0.020101	18.7
6	11	0.010575	9.9
7	5	0.005722	5.3
8	4	0.003161	2.9
9	3	0.001774	1.6
10	2	0.001010	0.9
11	0	0.000578	0
12	1	0.000811	0.8
Total	930	1.0	929.4

8.4 Inferences

8.4.1 Point Estimation

Let $\bar{x}$ denote the mean of a random sample of n observations from a logarithmic series distribution with parameter θ. The maximum likelihood estimate $\widehat{\theta}$ of θ is the solution of the equation

$$\widehat{\theta} = \frac{\bar{x}\ln(1-\widehat{\theta})}{\bar{x}\ln(1-\widehat{\theta})-1},$$

which can be solved numerically for a given sample mean. Williamson and Bretherton (1964) tabulated the values of $\widehat{\theta}$ for $\bar{x}$ ranging from 1 to 50.

Patil (1962) derived an asymptotic expression for the variance of the MLE, and it is given by

$$\mathrm{Var}(\widehat{\theta}) = \frac{\theta^{\,2}}{n\mu_2},$$

where μ_2 denotes the variance of the logarithmic series distribution with parameter θ (see Section 8.2).

Patil and Bildikar (1966) considered the problem of minimum variance unbiased estimation. Wani (1975) compared the MLE and the minimum variance unbiased estimator (MVUE) numerically and concluded that there is no clear-cut criterion to choose between these estimators. It should be noted that the MVUE also can not be expressed in a closed form.

8.4.2 Interval Estimation

Let $X_1, \ldots, X_n$ be a random sample from a logarithmic series distribution with parameter θ. Let Z denote the sum of the X_i's, and let $f(z|\ n,\ \theta)$ denote the probability mass function of Z (see Section 8.5). For an observed value z_0 of Z, a $(1-\alpha)$ confidence interval is (θ_L, θ_U), where θ_L and θ_U satisfy

$$\sum_{k=z_0}^{\infty} f(k|n, \theta_L) = \frac{\alpha}{2},$$

and

$$\sum_{k=1}^{z_0} f(k|n, \theta_U) = \frac{\alpha}{2}$$

respectively. Wani (1975) tabulated the values of (θ_L, θ_U) for $n = 10$, 15, and 20, and the confidence level 0.95.

8.5 Properties and Results

1. Recurrence Relations:

$$P(X = k + 1) = \tfrac{k\theta}{k+1} P(X = k), \quad k = 1, 2, \ldots$$
$$P(X = k - 1) = \tfrac{k}{(k-1)\theta} P(X = k), \quad k = 2, 3, \ldots$$

2. Let $X_1, \ldots, X_n$ be independent random variables, each having a logarithmic series distribution with parameter θ. The probability mass function of the $Z = \sum_{i=1}^{n} X_i$ is given by

$$P(Z = k) = \frac{n!|S_k^{(n)}|\theta^k}{k![-\ln(1-\theta)]^n}, \quad k = n, n+1, \ldots$$

   where $S_k^{(n)}$ denotes the Stirling number of the first kind (Abramowitz and Stegun 1965, p. 824).

8.6 Random Number Generation

The following algorithm is based on the inverse method. That is, for a random uniform$(0, 1)$ number u, the algorithm searches for k such that $P(X \le k - 1) < u \le P(X \le k)$.

Input:

θ = parameter

ns = desired number of random numbers

Output:

$x(1), \ldots, x(ns)$ are random numbers from the Logarithmic Series(θ) distribution

Algorithm 8.6.1

Set pk = $-\theta/\ln(1-\theta)$
    rpk = pk
For j = 1 to ns
        Generate u from uniform(0, 1)

```
        k = 1
1       If u ≤ pk, go to 2
        u = u - pk
        pk = pk *θ*k/(k + 1)
        k = k + 1
        goto 1
2       x(j) = k
        pk = rpk
[end j loop]
```

8.7 A Computational Algorithm for Probabilities

For a given θ and k, $P(X = k)$ can be computed in a straightforward manner. To compute $P(X \le k)$, compute first $P(X = 1)$, compute other probabilities recursively using the recurrence relation

$$P(X = i + 1) = \frac{i\theta}{i+1} P(X = i), \quad i = 1, 2, ..., k - 1 \ldots$$

and then compute $P(X \le k) = P(X = 1) + \sum_{i=2}^{k} P(X = i)$.

The above method is used to obtain the following algorithm.

Algorithm 8.7.1

```
Input:
      k = the positive integer at which the cdf is to be evaluated
      t = the value of the parameter 'theta'
      a = -1/ln(1-t)
Output:
      cdf = P(X <= k| t)

Set p1 = t
      cdf = p1
For i = 1 to k
      p1 = p1*i*t/(i+1)
      cdf = cdf + p1
(end i loop)
cdf = cdf*a
```

Chapter 9

Continuous Uniform Distribution

9.1 Description

The probability density function of a continuous uniform random variable over an interval [a, b] is given by

$$f(x; a, b) = \frac{1}{b-a}, \quad a \leq x \leq b.$$

The cumulative distribution function is given by

$$F(x|a, b) = \frac{x-a}{b-a}, \quad a \leq x \leq b.$$

The uniform distribution with support $[a, b]$ is denoted by uniform(a, b). This distribution is also called the *rectangular* distribution because of the shape of its pdf. (see Figure 9.1).

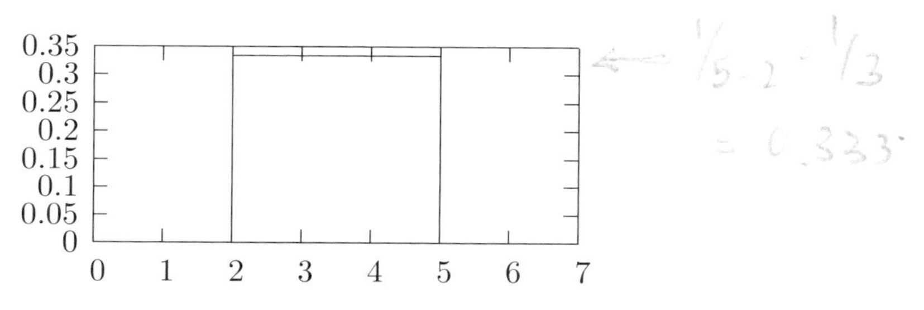

Figure 9.1 The pdf of Uniform(2,5)

9.2 Moments

Mean:	$\frac{b+a}{2}$
Variance:	$\frac{(b-a)^2}{12}$
Median:	$\frac{b+a}{2}$
Coefficient of Variation:	$\frac{(b-a)}{\sqrt{3}(b+a)}$
Mean Deviation:	$\frac{b-a}{4}$
Coefficient of Skewness:	0
Central Moments:	$\begin{cases} 0, & k = 1, 3, 5, \ldots, \\ \frac{(b-a)^k}{2^k(k+1)}, & k = 2, 4, 6, \ldots \end{cases}$
Moments about the Origin:	$E(X^k) = \frac{b^{k+1}-a^{k+1}}{(b-a)(k+1)}, \quad k = 1, 2, \cdots$
Moment Generating Function:	$\frac{e^{tb}-e^{ta}}{(b-a)t}$

9.3 Inferences

Let $X_1, \ldots, X_n$ be a random sample from a uniform(a, b) distribution. Let $X_{(1)}$ denote the smallest order statistic and $X_{(n)}$ denote the largest order statistic.

1. When b is known,

$$\widehat{a}_u = \frac{(n+1)X_{(1)} - b}{n}$$

is the uniformly minimum variance unbiased estimator (UMVUE) of a; if a is known, then

$$\widehat{b}_u = \frac{(n+1)X_{(n)} - a}{n}$$

is the UMVUE of b.

2. When both a and b are unknown,

$$\widehat{a} = \frac{nX_{(1)} - X_{(n)}}{n-1} \quad \text{and} \quad \widehat{b} = \frac{nX_{(n)} - X_{(1)}}{n-1}$$

are the UMVUEs of a and b, respectively.

9.4 Properties and Results

1. *Probability Integral Transformation:* Let X be a continuous random variable with cumulative distribution function $F(x)$. Then,

$$U = F(X) \sim \text{uniform}(0, 1).$$

2. Let X be a uniform(0, 1) random variable. Then,

$$-2\ln(X) \sim \chi^2_2.$$

3. Let $X_1, \ldots, X_n$ be independent uniform(0,1) random variables, and let $X_{(k)}$ denote the kth order statistic. Then $X_{(k)}$ follows a beta(k, $n - k + 1$) distribution.

4. Relation to Normal: See Section 10.11.

9.5 Random Number Generation

Uniform(0, 1) random variates generator is usually available as a built-in intrinsic function in many commonly used programming languages such as Fortran and C. To generate random numbers from uniform(a, b), use the result that if $U \sim$ uniform(0,1), then $X = a + U * (b - a) \sim \text{uniform}(a, b)$.

Chapter 10

Normal Distribution

10.1 Description

The probability density function of a normal random variable X with mean μ and standard deviation σ is given by

$$f(x|\mu,\sigma) = \frac{1}{\sigma\sqrt{2\pi}} \exp\left[-\frac{(x-\mu)^2}{2\sigma^2}\right], -\infty < x < \infty, \ -\infty < \mu < \infty, \sigma > 0.$$

This distribution is commonly denoted by $N(\mu, \sigma^2)$. The cumulative distribution function is given by

$$F(x|\mu,\sigma) = \int_{-\infty}^{x} f(t|\mu,\sigma)dt.$$

The normal random variable with mean $\mu = 0$ and standard deviation $\sigma = 1$ is called the standard normal random variable, and its cdf is denoted by $\Phi(z)$.

If X is a normal random variable with mean μ and standard deviation σ, then

$$P(X \le x) = P\left(Z \le \frac{x-\mu}{\sigma}\right) = \int_{-\infty}^{(x-\mu)/\sigma} \exp(-t^2/2)dt = \Phi\left(\frac{x-\mu}{\sigma}\right).$$

The mean μ is the location parameter, and the standard deviation σ is the scale parameter. See Figures 10.2 and 10.3.

The normal distribution is the most commonly used distribution to model univariate data from a population or from an experiment. In the following example, we illustrate a method of checking whether a sample is from a normal population.

An Example for Model Checking

Example 10.1.1 (Assessing Normality) Industrial hygiene is an important problem in situations where the employees are constantly exposed to workplace contaminants. In order to assess the exposure levels, hygienists monitor employees periodically. The following data represent the exposure measurements from a sample of 15 employees who were exposed to a chemical over a period of three months.

x : 69 75 82 93 98 102 54 59 104 63 67 66 89 79 77

We want to test whether the exposure data are from a normal distribution. Following the steps of Section 1.4.1, we first order the data. The ordered data $x_{(j)}$'s are given in the second column of Table 10.1. The cumulative probability level of $x_{(j)}$ is approximately equal to $(j - 0.5)/n$, where n is the number of data points. For these cumulative probabilities, standard normal quantiles are computed, and they are given in the fourth column of Table 10.1. For example, when $j = 4$, the observed 0.233333th quantile is 66 and the corresponding standard normal quantile $z_{(j)}$ is -0.7279. To compute the standard normal quantile for the 4th observation, select [StatCalc→Continuous→Normal→Probabilities, Percentiles and Moments], enter 0 for [Mean], 1 for [Std Dev], and 0.233333 for P(X<=x); click on [x] to get -0.7279. If the data are from a normal population, then the pairs $(x_{(j)}, z_{(j)})$ will be approximately linearly related. The plot of the pairs (Q–Q plot) is given in Figure 10.1. The Q–Q plot is nearly a line suggesting that the data are from a normal population.

If a graphical technique does not give a clear-cut result, a rigorous test, such as Shapiro–Wilk test and the correlation test, can be used. We shall use the test based on the correlation coefficient to test the normality of the exposure data. The correlation coefficient between the $x_{(j)}$'s and the $z_{(j)}$'s is 0.984. At the level 0.05, the critical value for $n = 15$ is 0.937 (see Looney and Gulledge 1985). Since the observed correlation coefficient is larger than the critical value, we have further evidence for our earlier conclusion that the data are from a normal population.

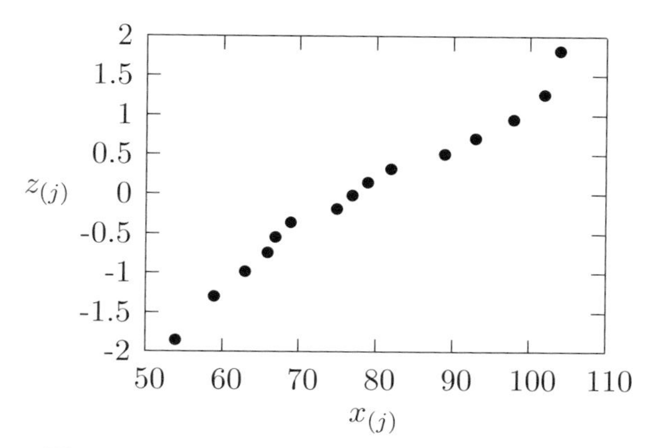

Figure 10.1 Q–Q plot of the Exposure Data

Table 10.1 Observed and normal quantiles for exposure data

j	Observed Quantiles $x_{(j)}$	Cumulative Probability Levels $(j-0.5)/15$	Standard Normal Quantile $z_{(j)}$
1	54	0.0333	-1.8339
2	59	0.1000	-1.2816
3	63	0.1667	-0.9674
4	66	0.2333	-0.7279
5	67	0.3000	-0.5244
6	69	0.3667	-0.3407
7	75	0.4333	-0.1679
8	77	0.5000	0.0000
9	79	0.5667	0.1679
10	82	0.6333	0.3407
11	89	0.7000	0.5244
12	93	0.7667	0.7279
13	98	0.8333	0.9674
14	102	0.9000	1.2816
15	104	0.9667	1.8339

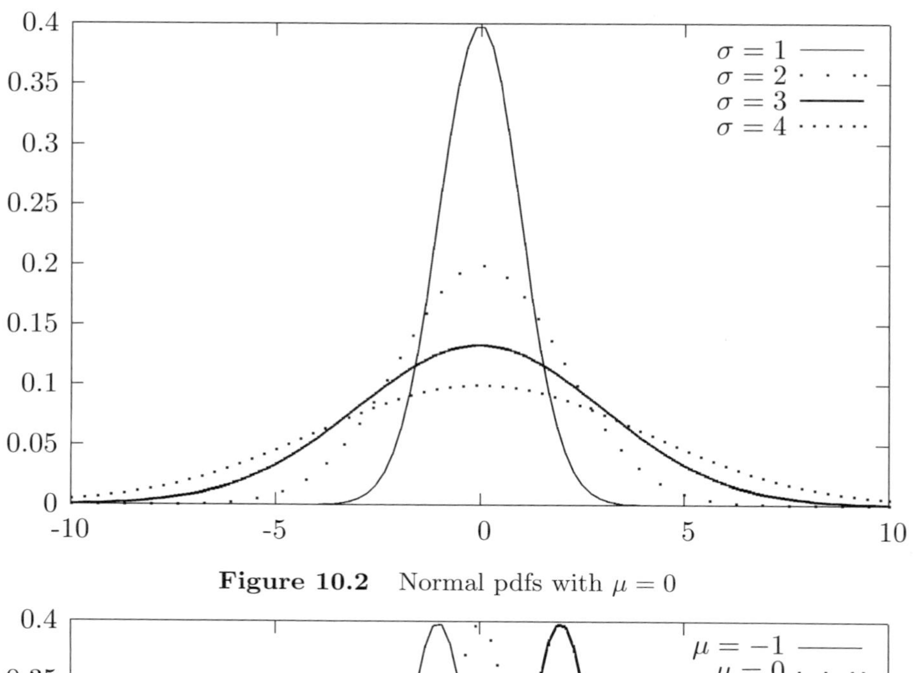

Figure 10.2 Normal pdfs with $\mu = 0$

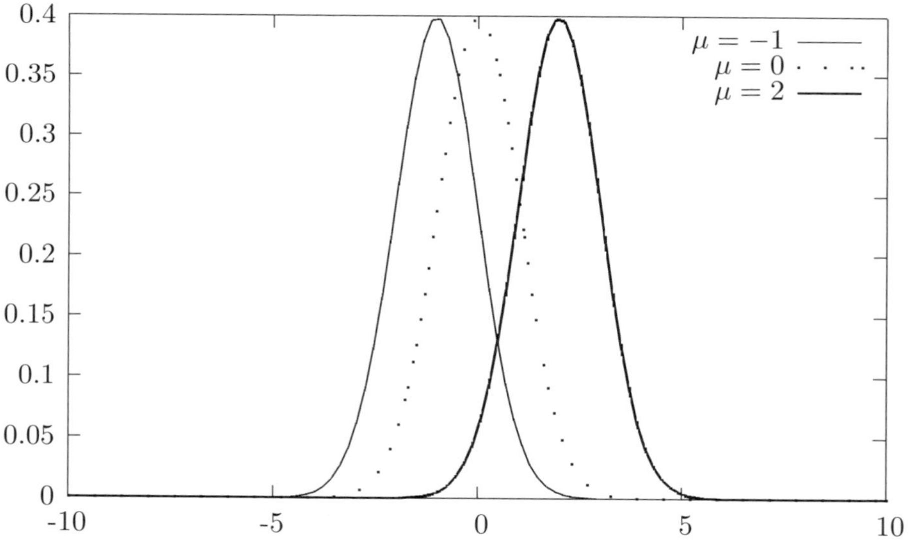

Figure 10.3 Normal pdfs with $\sigma = 1$

10.2 Moments

Mean:	μ
Variance:	σ^2
Coefficient of Variation:	σ/μ
Median:	μ
Mean Deviation:	$\sqrt{\frac{2\sigma^2}{\pi}}$
Coefficient Skewness:	0
Coefficient of Kurtosis:	3
Moments about the Origin:	$\begin{cases} \sigma^k \sum_{i=1}^{(k+1)/2} \frac{k!\mu^{2i-1}}{(2i-1)![(k+1)/2-i]!2^{(k+1)/2-i}\sigma^{2i-1}}, \\ \qquad k = 1, 3, 5, \ldots \\ \sigma^k \sum_{i=0}^{k/2} \frac{k!\mu^{2i}}{(2i)!(k/2-i)!2^{k/2-i}\sigma^{2i}}, \\ \qquad k = 2, 4, 6, \ldots \end{cases}$
Moments about the Mean:	$\begin{cases} 0, \quad k = 1, 3, 5, \ldots, \\ \frac{k!}{2^{k/2}(k/2)!}\sigma^k, \quad k = 2, 4, 6, \ldots \end{cases}$
Moment Generating Function:	$E(e^{tx}) = \exp\left(t\mu + t^2\sigma^2/2\right)$
A Recurrence Relation:	$E(X^k) = (k-1)\sigma^2 E(X^{k-2}) + \mu E(X^{k-1}),$ $k = 3, 4, \cdots$

10.3 Computing Table Values

The dialog box [StatCalc→Continuous→Normal] computes the following table values and other statistics.

1. Tail Probabilities, Percentiles, and Moments.
2. Test and Confidence Interval for the Mean [Section 10.4].
3. Power of the t-test [Section 10.4].

4. Test and Confidence Interval for the Variance [Section 10.4].
5. Test and Confidence Interval for the Variance Ratio [Section 10.5].
6. Two-Sample t-test and Confidence Interval [Section 10.5].
7. Two-Sample Test with No Assumption about the Variances [Section 10.5].
8. Power of the Two-Sample t-test [Section 10.5].
9. Tolerance Intervals for Normal Distribution [Section 10.6].
10. Tolerance Intervals Controlling both Tails [Section 10.6].
11. Simultaneous Tests for Quantiles [Section 10.6].

The dialog box [StatCalc→Continuous→Normal→Probabilities, Percentiles and Moments] computes the tail probabilities, critical points, parameters, and moments.

To compute probabilities: Enter the values of the mean, standard deviation, and the value of x at which the cdf is to be computed; click [P(X <= x)] radio button.

Example 10.3.1 When mean = 1.0, standard deviation = 2.0, and the value $x = 3.5$, $P(X \leq 3.5) = 0.89435$ and $P(X > 3.5) = 0.10565$.

To compute percentiles: Enter the values of the mean, standard deviation, and the cumulative probability P(X <= x); click on [x] radio button.

Example 10.3.2 When mean = 1.0, standard deviation = 2.0, and the cumulative probability P(X <= x) = 0.95, the 95th percentile is 4.28971. That is,

$$P(X \leq 4.28971) = 0.95.$$

To compute the mean: Enter the values of the standard deviation, x, and P(X <= x). Click [Mean].

Example 10.3.3 When standard deviation = 3, $x = 3.5$, and P(X <= x) = 0.97, the value of the mean is -2.14238.

To compute the standard deviation: Enter the values of the mean, x, and P(X <= x). Click [Std Dev].

Example 10.3.4 When mean = 3, $x = 3.5$, and P(X <= x) = 0.97, the standard deviation is 0.265845.

To compute moments: Enter the values of the mean and standard deviation; click [M] button.

Some Illustrative Examples

Example 10.3.5 An electric bulb manufacturer reports that the average life-span of 100W bulbs is 1100 h with a standard deviation of 100 h. Assume that the life hours distribution is normal.

a. Find the percentage of bulbs that will last at least 1000 h.
b. Find the percentage of bulbs with lifetime between 900 and 1200 h.
c. Find the 90th percentile of the life hours.

Solution: Select the dialog box [StatCalc→Continuous→Normal→Probabilities, Percentiles and Moments].

a. To find the percentage, enter 1100 for the mean, 100 for the standard deviation, and 1000 for the observed x; click [P(X <= x)] radio button to get $\mathrm{P(X \le 1000) = 0.1587}$ and $\mathrm{P(X > 1000) = 0.8413}$. That is, about 84% of the bulbs will last more than 1000 h.

b.

$$\begin{aligned} P(900 \le X \le 1200) &= P(X \le 1200) - P(X \le 900) \\ &= 0.841345 - 0.022750 \\ &= 0.818595. \end{aligned}$$

That is, about 82% of the bulbs will last between 900 and 1200 h.

c. To find the 90th percentile, enter 1100 for the mean, 100 for the standard deviation, and 0.90 for the cumulative probability; click on [x] to get 1228.16. That is, 90% of the bulbs will last less than 1228 h; and 10% of the bulbs will last more than 1228 h.

Example 10.3.6 Suppose that the weekly demand for 5-lb sacks of onions at a grocery store is normally distributed with mean 140 sacks and standard deviation 10.

a. If the store stocks 160 sacks every week, find the percentage of weeks that the store has overstocked onions.
b. How many sacks should the store keep in stock each week in order to meet the demand for 95% of the weeks?

Solution:

a. Let X denote the weekly demand. We need to find percentage of the weeks the demand is less than the stock. Enter 140 for the mean, 10 for the standard deviation, and click [P(X <= x)] radio button to get $P(X \le 160) = 0.97725$. This probability means that about 98% of the weeks the demand will be less than the supply.

b. Here, we need to find the 95th percentile of the normal distribution; that is, the value of x such that $P(X \leq x) = 0.95$. Using *StatCalc*, we get the value of $x = 156.449$. This means that the store has to stock 157 sacks each week in order to meet the demand for 95% of the weeks.

Example 10.3.7 A machine is set to pack 3-lb of ground beef per package. Over a long period of time, it was found that the average amount packed was 3 lb with a standard deviation of 0.1 lb. Assume that the weights of the packages are normally distributed.

a. Find the percentage of packages weighing more than 3.1 lb.

b. At what level, should the machine be set, so that no more than 5% of the packages weigh less than 2.9 lb?

Solution: Let X be the actual weight of a randomly selected package. Then, X is normally distributed with mean 3 lb and standard deviation 0.1 lb.

a. To find the percentage, enter 3 for the mean, 0.1 for the standard deviation, and 3.1 for the x; click [P(X <= x)] radio button to get P(X > 3.1) = 0.158655. That is, about 16% of the packages will weigh more than 3.1 lb.

b. We are looking for the value of the mean μ such that $P(X < 2.9) = 0.05$. To get the value of the mean, enter 0.1 for the standard deviation, 2.9 for x, 0.05 for $P(X \leq x)$, and then click on [Mean] to get 3.06449. That is, the machine needs to be set at about 3.07 pounds so that fewer than 5% of the packages weigh less than 2.9 lb.

Example 10.3.8 A manufacturing company received a large quantity of bolts from one of its suppliers. A bolt is useable if it is 3.9 to 4.1 in. long. Inspection of a sample of 50 bolts revealed that the average length is 3.95 in. with standard deviation 0.1 in. Assume that the distribution of lengths is normal.

a. Find an approximate proportion of bolts is useable.

b. Find an approximate proportion of bolts that are longer than 4.1 in.

c. Find an approximate 95th percentile of the lengths of all bolts.

Solution: Assume that the lengths of bolts form a normal population with the mean μ and standard deviation σ. If μ and σ are known, then exact proportions and percentile can be computed using *StatCalc*. Since they are unknown, we can use the sample mean and standard deviation to find approximate solutions to the problem.

a. The proportion of bolts useable is given by $P(3.9 \le X \le 4.1) = P(X \le 4.1) - P(X \le 3.9)$, where X is a normal random variable with mean 3.95 and standard deviation 0.1. Using *StatCalc*, we get $P(X \le 4.1) - P(X \le 3.9) = 0.933193 - 0.308538 = 0.624655$. Thus, about 62% of bolts are useable.

b. This proportion is given by $P(X \ge 4.1) = 1 - P(X \le 4.1) = 1 - 0.933193 = 0.0668074$. Thus, about 7% of bolts are longer than 4.1 inch.

c. To find an approximate 95th percentile, enter 3.95 for the mean, 0.1 for the standard deviation, 0.95 for the probability [P(X <= x)], and click [x] to get 4.11449. This means that approximately 95% of the bolts are shorter than 4.11449 inch.

10.4 One-Sample Inference

10.4.1 Point Estimation

Let $X_1, \ldots, X_n$ be a random sample from a normal population with mean μ and standard deviation σ. The sample mean

$$\bar{X} = \frac{1}{n}\sum_{i=1}^{n} X_i,$$

and the sample variance

$$S^2 = \frac{1}{n-1}\sum_{i=1}^{n}(X_i - \bar{X})^2$$

are the uniformly minimum variance unbiased estimators of μ and σ^2, respectively. The sample mean is the maximum likelihood estimator of μ; however, the maximum likelihood estimator of σ^2 is $(n-1)S^2/n$.

10.4.2 Test for the Mean and Power Computation

Hypothesis Test about the Mean (t-test)

The test statistic for testing null hypothesis $H_0 : \mu = \mu_0$ is given by

$$t = \frac{\bar{X} - \mu_0}{S/\sqrt{n}}, \tag{10.4.1}$$

where t follows a t distribution with df $= n - 1$.

Let $(\bar{x}, s^2)$ be an observed value of $(\bar{X}, S^2)$. Then $t_0 = \frac{\bar{x}-\mu_0}{s/\sqrt{n}}$ is the observed value of t in (10.4.1). For a given level α, the null hypothesis $H_0 : \mu = \mu_0$ will be rejected in favor of

$$H_a : \mu \neq \mu_0 \text{ if the p-value } P(|t| > |t_0|) < \alpha,$$

for testing $H_0 : \mu \geq \mu_0$ vs. $H_a : \mu < \mu_0$, the H_0 will be rejected if the

$$\text{p-value} = P(t \leq t_0) < \alpha,$$

and for testing $H_0 : \mu \leq \mu_0$ vs. $H_a : \mu > \mu_0$, the H_0 will be rejected if the

$$\text{p-value} = P(t \geq t_0) < \alpha.$$

Power Computation

Consider the hypotheses

$$H_0 : \mu \leq \mu_0 \quad \text{vs.} \quad H_a : \mu > \mu_0.$$

For a given nominal level α, the power of the t-test is the probability of rejecting the null hypothesis when the true mean μ is indeed greater than μ_0, and is given by

$$P(t > t_{n-1,1-\alpha}|H_a) = P(t_{n-1}(\delta) > t_{n-1,1-\alpha}), \tag{10.4.2}$$

where t is given in (10.4.1), $t_{n-1,1-\alpha}$ denotes the $(1 - \alpha)$th quantile of the t-distribution with degrees of freedom $n - 1$, and $t_{n-1}(\delta)$ denotes the noncentral t random variable with degrees of freedom $n - 1$ and the noncentrality parameter

$$\delta = \frac{\sqrt{n}(\mu - \mu_0)}{\sigma}.$$

The power of a two-tail test is similarly calculated. For a given μ, μ_0, σ and level α, *StatCalc* computes the power using (10.4.2).

To compute p-values for hypothesis testing about μ: Select [StatCalc→Continuous →Normal→t-test and CI for Mean] from *StatCalc*, enter the values of the sample mean, sample standard deviation, sample size [S Size n], and the value of the mean under the null hypothesis. Click [p-values for] to get the p-values for various alternative hypotheses.

Example 10.4.1 (Hypothesis Testing) Suppose that a sample of 20 observations from a normal population produced a mean of 3.4 and a standard deviation of 2.1. Consider testing

$$H_0 : \mu \le 2.5 \text{ vs. } H_a : \mu > 2.5.$$

To compute the p-value for testing above hypotheses, select [StatCalc→Continuous →Normal→t-test and CI for Mean] from *StatCalc*, enter 3.4 for the sample mean, 2.1 for the sample standard deviation, 20 for the sample size, and 2.5 for [H0:M = M0] and click [p-values for] to get 0.0352254. That is, the p-value for testing above hypotheses is 0.0352254. Thus, at 5% level, we have enough evidence to conclude that the true population mean is greater than 2.5.

Furthermore, note that for the two-sided hypothesis, that is, when

$$H_0 : \mu = 2.5 \text{ vs. } H_a : \mu \ne 2.5,$$

the p-value is 0.0704508. Now, the null hypothesis cannot be rejected at the level of significance 0.05. The value of the t-test statistic for this problem is 1.91663.

Sample Size for One-Sample t-test: For a given level of significance, the true mean and hypothesized mean of a normal population, and the standard deviation, the dialog box [StatCalc→Continuous → Normal → Sample Size for t-test] computes the sample size that is required to have a specified power. To compute the sample size, enter the hypothesized value of the population mean in [H0: M = M0], the population mean in [Population M], population standard deviation, level of the test and power. Click [Sample Size for].

Example 10.4.2 (Sample Size Calculation) An experimenter believes that the actual mean of the population under study is 1 unit more than the hypothesized value $\mu_0 = 3$. From the past study, he learned that the population standard deviation is 1.3. He decides to use one-sample t-test, and wants to determine the sample size to attain a power of 0.90 at the level 0.05. The hypotheses for his study will be

$$H_0 : \mu \le 3 \text{ vs. } H_a\colon \mu > 3.$$

To compute the required sample size using *StatCalc*, enter 3 for [H0: M = M0], 4 for the population mean, 1.3 for the population standard deviation, 0.05 for the level, and 0.9 for the power; click [Sample Size for] to get 16. Thus, a sample of 16 observations will be sufficient to detect the difference between the true mean and hypothesized mean with a power of 90%.

10.4.3 Interval Estimation for the Mean

Confidence Interval for μ (t-interval): A $1-\alpha$ confidence interval for the mean μ is given by

$$\bar{X} \pm t_{n-1,\,1-\alpha/2}\frac{S}{\sqrt{n}},$$

where $t_{n-1,1-\alpha/2}$ is the $(1-\alpha/2)$th quantile of a t distribution with df $= n-1$.

Prediction Interval for an Individual: A $1-\alpha$ prediction interval for an individual (from the normal population from which the sample was drawn) is given by

$$\bar{X} \pm t_{n-1,\,1-\alpha/2} S\sqrt{1+1/n}.$$

For a given sample mean, sample standard deviation, and sample size, the dialog box [StatCalc→Continuous→Normal→t-test and CI for Mean] computes the p-values of the t-test and confidence intervals for the mean.

To compute a confidence interval for μ: Select the dialog box [StatCalc→ Continuous→Normal→t-test and CI for Mean], enter the values of the sample mean, sample standard deviation, sample size, and the confidence level. Click [1-sided] to get one-sided lower and upper limits for μ. Click [2-sided] to get confidence interval for μ.

Example 10.4.3 Let us compute a 95% confidence interval for the true population mean based on summary statistics given in Example 10.4.1. In the dialog box [StatCalc→ Continuous→Normal→t-test and CI for Mean], enter 3.4 for the sample mean, 2.1 for the sample standard deviation, 20 for the sample size, and 0.95 for the confidence level. Click [1-sided] to get 2.58804 and 4.21196. These are the one-sided limits for μ. That is, the interval (2.58804, ∞) would contain the population mean μ with 95% confidence. The interval (-∞, 4.21196) would contain the population mean μ with 95% confidence. To get a two-sided confidence interval for μ, click on [2-sided] to get 2.41717 and 4.38283. This means that the interval (2.41717, 4.38283) would contain the true mean with 95% confidence.

Some Illustrative Examples

The following examples illustrate the one-sample inferential procedures for a normal mean.

Example 10.4.4 A marketing agency wants to estimate the average annual income of all households in a suburban area of a large city. A random sample of 40 households from the area yielded a sample mean of $65,000 with standard deviation $4,500. Assume that the incomes follow a normal distribution.

a. Construct a 95% confidence interval for the true mean income of all the households in the suburb community.

b. Do these summary statistics indicate that the true mean income is greater than $63,000?

Solution:

a. To construct a 95% confidence interval for the true mean income, enter 65000 for the sample mean, 4500 for the sample standard deviation, and 40 for the sample size, and 0.95 for the confidence level. Click [2-sided] to get 63560.8 and 66439.2. That is, the actual mean income is somewhere between $63,560.80 and $66439.2 with 95% confidence.

b. It is clear from part **a** that the mean income is greater than $63,000. However, to understand the significance of the result, we formulate the following hypothesis testing problem. Let μ denote the true mean income. We want to test

$$H_0 : \mu \leq 63000 \quad \text{vs.} \quad H_a : \mu > 63000.$$

To compute the p-value for the above test, enter the sample mean, standard deviation and the sample size as in part **a**, and 63000 for [Ha: M = M0]. Click [p-values for] to get 0.00384435. Since this p-value is less than any practical level of significance, the summary statistics provide strong evidence to indicate that the mean income is greater than $63,000.

Example 10.4.5 A light bulb manufacturer considering a new method that is supposed to increase the average lifetime of bulbs by at least 100 h. The mean and standard deviation of the life hours of bulbs produced by the existing method are 1200 and 140 h respectively. The manufacturer decides to test if the new method really increases the mean life hour of the bulbs. How many new bulbs should he test so that the test will have a power of 0.90 at the level of significance 0.05?

Solution: Let μ denote the actual mean life hours of the bulbs manufactured using the new method. The hypotheses of interest here are

$$H_0 : \mu \leq 1200 \quad \text{vs.} \quad H_a : \mu > 1200.$$

Enter 1200 for [H0: M = M0], 1300 for the population mean, 140 for the population standard deviation (it is assumed that the standard deviations of the existing method and old method are the same), 0.05 for the level and 0.9 for the power. Click [Sample Sizes for] to get 19. Thus, nineteen bulbs should be manufactured and tested to check if the new method would increase the average life hours of the bulbs.

10.4.4 Test and Interval Estimation for the Variance

Let S^2 denote the variance of a sample of n observations from a normal population with mean μ and variance σ^2. The pivotal quantity for testing and interval estimation of a normal variance is given by

$$Q = \frac{(n-1)S^2}{\sigma^2},$$

which follows a chi-square distribution with df $= n - 1$.

Test about a Normal Variance

Let Q_0 be an observed value of Q. For testing

$$H_0 : \sigma^2 = \sigma_0^2 \quad \text{vs.} \quad H_a : \sigma^2 \neq \sigma_0^2,$$

a size α test rejects H_0 if $2\min\{P(\chi^2_{n-1} > Q_0), P(\chi^2_{n-1} < Q_0)\} < \alpha$. For testing

$$H_0 : \sigma^2 \leq \sigma_0^2 \quad \text{vs.} \quad H_a : \sigma^2 > \sigma_0^2,$$

the null hypothesis H_0 will be rejected if $P(\chi^2_{n-1} > Q_0) < \alpha$, and for testing

$$H_0 : \sigma^2 \geq \sigma_0^2 \quad \text{vs.} \quad H_a : \sigma^2 < \sigma_0^2,$$

the null hypothesis H_0 will be rejected if $P(\chi^2_{n-1} < Q_0) < \alpha$.

Confidence Interval for a Normal Variance

A $1 - \alpha$ confidence interval for the variance σ^2 is given by

$$\left(\frac{(n-1)S^2}{\chi^2_{n-1,1-\alpha/2}}, \frac{(n-1)S^2}{\chi^2_{n-1,\alpha/2}} \right),$$

where $\chi^2_{m,p}$ denotes the pth quantile of a chi-square distribution with df $= m$.

For a given sample variance and sample size, the dialog box [StatCalc → Continuous → Normal → t-Test and CI for the Variance] computes the confidence interval for the population variance σ^2, and p-values for hypothesis testing about σ^2.

To compute a confidence interval for σ^2: Enter the value of the sample size, sample variance, and the confidence level. Click [1-sided] to get one-sided lower and upper limits for σ^2. Click [2-sided] to get confidence interval for σ^2.

Example 10.4.6 Suppose that a sample of 20 observations from a normal population produced a variance of 12. To compute a 90% confidence interval for σ^2, enter 20 for [Sample Size], 12 for [Sample Variance], and 0.90 for [Confidence Level]. Click [1-sided] to get 8.38125 and 19.5693. These are the one-sided limits for σ^2. That is, the interval (8.38125, ∞) would contain the population variance σ^2 with 90% confidence. The interval (0, 19.5693) would contain the population variance σ^2 with 90% confidence. To get a two-sided confidence interval for σ^2, click on [2-sided] to get 7.56381 and 22.5363. This means that the interval (7.56381, 22.5363) would contain σ^2 with 90% confidence.

To compute p-values for hypothesis testing about σ^2: Enter the summary statistics as in the above example, and the specified value of σ^2 under the null hypothesis. Click [p-values for] to get the p-values for various alternative hypotheses.

Example 10.4.7 Suppose we want to test

$$H_0 : \sigma^2 \leq 9 \quad \text{vs.} \quad H_a : \sigma^2 > 9 \qquad (10.4.3)$$

at the level of 0.05 using the summary statistics given in Example 10.4.6. After entering the summary statistics, enter 9 for [H0: V = V0]. Click [p-values for] to get 0.14986. Since this p-value is not less than 0.05, the null hypothesis in (10.4.3) can not be rejected at the level of significance 0.05. We conclude that the summary statistics do not provide sufficient evidence to indicate that the true population variance is greater than 9.

The examples given below illustrate the inferential procedures about a normal variance.

Example 10.4.8 (*Test about* σ^2) A hardware manufacturer was asked to produce a batch of 3-in. screws with a specification that the standard deviation of the lengths of all the screws should not exceed 0.1 in. At the end of a day's production, a sample of 27 screws was measured, and the sample standard deviation was calculated as 0.09. Does this sample standard deviation indicate that the actual standard deviation of all the screws produced during that day is less than 0.1 in.?

Solution: Let σ denote the standard deviation of all the screws produced during that day. The appropriate hypotheses for the problem are

$$H_0 : \sigma \geq 0.1 \;\; \text{vs.} \;\; H_a : \sigma < 0.1 \iff H_0 : \sigma^2 \geq 0.01 \;\; \text{vs.} \;\; H_a : \sigma^2 < 0.01.$$

Note that the sample variance is $(0.09)^2 = 0.0081$. To compute the p-value for the above test, enter 27 for the sample size, 0.0081 for the sample variance, 0.01 for [H0: V = V0], and click on [p-values for]. The computed p-value is 0.26114. Since this p-value is not smaller than any practical level of significance, we can not conclude that the standard deviation of all the screws made during that day is less than 0.1 in.

Example 10.4.9 (*Confidence Interval for* σ^2) An agricultural student wants to estimate the variance of the yields of tomato plants. He selected a sample of 18 plants for the study. After the harvest, he found that the mean yield was 38.5 tomatoes with standard deviation 3.4. Assuming a normal model for the yields of tomato, construct a 90% confidence interval for the variance of the yields of all tomato plants.

Solution: To construct a 90% confidence interval for the variance, enter 18 for the sample size, $(3.4)^2 = 11.56$ for the sample variance, and 0.90 for the confidence level. Click [2-sided] to get 7.12362 and 22.6621. Thus, the true variance of tomato yields is somewhere between 7.12 and 22.66 with 90% confidence.

10.5 Two-Sample Inference

Let S_i^2 denote the variance of a random sample of n_i observations from $N(\mu_i, \sigma_i^2)$, $i = 1, 2$. The following inferential procedures for the ratio σ_1^2/σ_2^2 are based on the F statistic given by

$$F = \frac{S_1^2}{S_2^2}. \tag{10.5.1}$$

10.5.1 Inference for the Ratio of Variances

Hypothesis Test for the Ratio of Variances

Consider testing $H_0 : \sigma_1^2/\sigma_2^2 = 1$. When H_0 is true, then the F statistic in (10.5.1) follows an F_{n_1-1,n_2-1} distribution. Let F_0 be an observed value of the F in (10.5.1). A size α test rejects the null hypothesis in favor of

$$H_a : \sigma_1^2 > \sigma_2^2 \text{ if the p-value } P\left(F_{n_1-1,n_2-1} > F_0\right) < \alpha.$$

For testing $H_0 : \sigma_1^2 \geq \sigma_2^2$ vs. $H_a : \sigma_1^2 < \sigma_2^2$, the null hypothesis will be rejected if the p-value $P\left(F_{n_1-1,n_2-1} < F_0\right) < \alpha$. For a two-tail test, the null hypothesis $H_0 : \sigma_1^2 = \sigma_2^2$ will be rejected if either tail p-value is less than $\alpha/2$.

Interval Estimation for the Ratio of Variances

A $1-\alpha$ confidence interval is given by

$$\left(\frac{S_1^2}{S_2^2}F_{n_2-1,n_1-1,\alpha/2}, \frac{S_1^2}{S_2^2}F_{n_2-1,n_1-1,1-\alpha/2}\right),$$

where $F_{m,n,p}$ denotes the pth quantile of an F distribution with the numerator df $= m$ and the denominator df $= n$. The above confidence interval can be obtained from the distributional result that $\left(\frac{S_2^2\sigma_1^2}{S_1^2\sigma_2^2}\right) \sim F_{n_2-1,n_1-1}$.

The dialog box [StatCalc→Continuous → Normal → Test and CI for the Variance Ratio] computes confidence intervals as well as the p-values for testing the ratio of two variances.

To compute a confidence interval for σ_1^2/σ_2^2: Enter the values of the sample sizes and sample variances, and the confidence level. Click [1-sided] to get one-sided lower and upper limits for σ_1^2/σ_2^2. Click [2-sided] to get confidence interval for σ_1^2/σ_2^2.

Example 10.5.1 (CI for σ_1^2/σ_2^2*)* A sample of 8 observations from a normal population produced a variance of 4.41. A sample of 11 observations from another normal population yielded a variance of 2.89. To compute a 95% confidence interval for σ_1^2/σ_2^2, select the dialog box [StatCalc→Continuous → Normal → Test and CI for the Variance Ratio], enter the sample sizes, sample variances, and 0.95 for the confidence level. Click [1-sided] to get 0.4196 and 4.7846. This means that the interval (0.4196, ∞) would contain the ratio σ_1^2/σ_2^2 with 95% confidence.

Furthermore, we can conclude that the interval (0, 4.7846) would contain the variance ratio with 95% confidence. Click [2-sided] to get 0.3863 and 7.2652. This means that the interval (0.3863, 7.2652) would contain the variance ratio with 95% confidence.

Example 10.5.2 (Hypothesis Tests for σ_1^2/σ_2^2) Suppose we want to test

$$H_0 : \sigma_1^2 = \sigma_2^2 \text{ vs. } H_a : \sigma_1^2 \neq \sigma_2^2,$$

at the level of 0.05 using the summary statistics given in Example 10.5.1. To compute the p-value, enter the summary statistics in the dialog box, and click on [p-values for] to get 0.5251. Since the p-value is greater than 0.05, we can not conclude that the population variances are significantly different.

There are two procedures available to make inference about the mean difference $\mu_1 - \mu_2$; one is based on the assumption that $\sigma_1^2 = \sigma_2^2$ and another is based on no assumption about the variances. In practice, the equality of the variances is tested first using the F test given above. If the assumption of equality of variances is tenable, then we use the two-sample t procedures (see Section 10.5.2) for making inference about $\mu_1 - \mu_2$; otherwise, we use the approximate procedure (see Section 10.5.3) known as Welch's *approximate degrees of freedom* method.

Remark 10.5.1 The above approach of selecting a two-sample test is criticized by many authors (see Moser and Stevens 1992 and Zimmerman 2004). In general, many authors suggested using the Welch test when the variances are unknown. Nevertheless, for the sake of completeness and illustrative purpose, we consider both approaches in the sequel.

10.5.2 Inference for the Difference between Two Means when the Variances are Equal

Let $\bar{X}_i$ and S_i^2 denote, respectively, the mean and variance of a random sample of n_i observations from $N(\mu_i, \sigma_i^2)$, $i = 1, 2$. Let

$$S_p^2 = \frac{(n_1 - 1)S_1^2 + (n_2 - 1)S_2^2}{n_1 + n_2 - 2}. \quad (10.5.2)$$

The following inferential procedures for $\mu_1 - \mu_2$ are based on the sample means and the pooled variance S_p^2, and are valid only when $\sigma_1^2 = \sigma_2^2$.

Two-Sample t Test

The test statistic for testing $H_0: \mu_1 = \mu_2$ is given by

$$t_2 = \frac{(\bar{X}_1 - \bar{X}_2)}{\sqrt{S_p^2(1/n_1 + 1/n_2)}} \tag{10.5.3}$$

which follows a t-distribution with degrees of freedom $n_1 + n_2 - 2$ provided $\sigma_1^2 = \sigma_2^2$. Let t_{20} be an observed value of t_2. For a given level α, the null hypothesis will be rejected in favor of

$$H_a: \mu_1 \neq \mu_2 \text{ if the p-value } P(|t_2| > |t_{20}|) < \alpha,$$

in favor of

$$H_a: \mu_1 < \mu_2 \text{ if the p-value } P(t_2 < t_{20}) < \alpha,$$

and in favor of

$$H_a: \mu_1 > \mu_2 \text{ if the p-value } P(t_2 > t_{20}) < \alpha.$$

Power of the Two-Sample t-test

Consider the hypotheses

$$H_0: \mu_1 \leq \mu_2 \quad \text{vs.} \quad H_a: \mu_1 > \mu_2.$$

For a given level α, the power of the two-sample t-test is the probability of rejecting the null hypothesis when μ_1 is indeed greater than μ_2, and is given by

$$P(t_2 > t_{n_1+n_2-2,1-\alpha}) = P(t_{n_1+n_2-2}(\delta) > t_{n_1+n_2-2,1-\alpha}), \tag{10.5.4}$$

where t_2 is given in (10.5.3), $t_{n_1+n_2-2,1-\alpha}$ denotes the $(1-\alpha)$th quantile of a t-distribution with degrees of freedom $n_1 + n_2 - 2$, $t_{n_1+n_2-2}(\delta)$ denotes the noncentral t random variable with the degrees of freedom $n_1 + n_2 - 2$ and noncentrality parameter

$$\delta = \frac{(\mu_1 - \mu_2)}{\sigma\sqrt{1/n_1 + 1/n_2}}.$$

The power of a two-tail test is similarly calculated. For given sample sizes, $\mu_1 - \mu_2$, common σ and the level of significance, *StatCalc* computes the power using (10.5.4).

Interval Estimation of $\mu_1 - \mu_2$

A $1-\alpha$ confidence interval based on the test-statistic in (10.5.3) is given by

$$\bar{X}_1 - \bar{X}_2 \pm t_{n_1+n_2-2,1-\alpha/2}\sqrt{S_p^2(1/n_1+1/n_2)},$$

where $t_{n_1+n_2-2,1-\alpha/2}$ denotes the $(1-\alpha/2)$th quantile of a t-distribution with n_1+n_2-2 degrees of freedom. This confidence interval is valid only when $\sigma_1^2 = \sigma_2^2$.

The dialog box [StatCalc→Continuous → Normal → Two-Sample t-test and CI] computes the p-values for testing the difference between two normal means and confidence intervals for the difference between the means. The results are valid only when $\sigma_1^2 = \sigma_2^2$.

To compute a confidence interval for $\mu_1 - \mu_2$: Enter the values of the sample means, sample standard deviations, sample sizes, and the confidence level. Click [1-sided] to get one-sided lower and upper limits for $\mu_1 - \mu_2$. Click [2-sided] to get confidence interval for $\mu_1 - \mu_2$.

Example 10.5.3 (Test about σ_1^2/σ_2^2) A sample of 8 observations from a normal population with mean μ_1, and variance σ_1^2 produced a mean of 4 and standard deviation of 2.1. A sample of 11 observations from another normal population with mean μ_2, and variance σ_2^2 yielded a mean of 2 with standard deviation of 1.7. Since the inferential procedures given in this section are appropriate only when the population variances are equal, we first want to test that if the variances are indeed equal (see Section 10.5.1). The test for equality of variances yielded a p-value of 0.525154, and hence the assumption of equality of population variances seems to be tenable.

To compute a 95% confidence interval for $\mu_1 - \mu_2$ using *StatCalc*, enter the sample means, standard deviations, and the sample sizes, and 0.95 for the confidence level. Click [1-sided] to get 0.484333 and 3.51567. This means that the interval (0.48433, ∞) would contain the difference $\mu_1 - \mu_2$ with 95% confidence. Furthermore, we can conclude that the interval $(-\infty, 3.51567)$ would contain the difference $\mu_1 - \mu_2$ with 95% confidence. Click [2-sided] to get 0.161782 and 3.83822. This means that the interval (0.161782, 3.83822) would contain the difference $\mu_1 - \mu_2$ with 95% confidence.

To compute p-values for testing $\mu_1 - \mu_2$: Select the dialog box [StatCalc→ Continuous → Normal → Two-Sample t-test and CI], enter the values of the sample sizes, sample means and sample standard deviations, and click [p-values for] to get the p-values for a right-tail test, left-tail test and two-tail test.

Example 10.5.4 Suppose we want to test

$$H_0 : \mu_1 \leq \mu_2 \text{ vs. } H_a : \mu_1 > \mu_2$$

at the level of 0.05 using the summary statistics given in Example 10.5.3. To compute the p-value, enter the summary statistics in the dialog box, and click on [p-values for] to get 0.0173486. Since the p-value is less than 0.05, we conclude that $\mu_1 > \mu_2$.

Power Calculation of Two-Sample t-test: The dialog box [StatCalc→Continuous → Normal → Two-Sample Case → Power Computation] computes the power of the two-sample t-test for

$$H_0 : \mu_1 \leq \mu_2 \text{ vs. } Ha : \mu_1 > \mu_2$$

when it is assumed that $\sigma_1^2 = \sigma_2^2$. To compute the power, enter the values of the level α of the test, the difference between the population means, the value of the common standard deviation σ and sample sizes n_1 and n_2. Click [Power]. Power of a two-tail test can be computed by entering $\alpha/2$ for the level.

Example 10.5.5 Suppose that the difference between two normal means is 1.5 and the common standard deviation is 2. It is desired to test

$$H_0 : \mu_1 \leq \mu_2 \text{ vs. } H_0 : \mu_1 > \mu_2$$

at the level of significance 0.05. To compute the power when each sample size is 27, enter 0.05 for level, 1.5 for the mean difference, 2 for the common σ, 27 for n_1 and n_2; click [Power] to get 0.858742.

Sample Size Calculation: In practical applications, it is usually desired to compute the sample sizes required to attain a given power. This can be done by a trial-error method. Suppose in the above example we need to determine the sample sizes required to have a power of 0.90. By trying a few sample sizes more than 27, we can find the required sample size as 32 from each population. In this case, the power is 0.906942. Also, note that when $n_1 = 27$ and $n_2 = 37$, the power is 0.900729.

The following examples illustrate the inferential procedures for the difference between two normal means.

Example 10.5.6 A company, which employs thousands of computer programmers, wants to compare the mean difference between the salaries of the male and female programmers. A sample of 23 male programmers, and a sample of 19 female programmers were selected, and programmers' salaries were recorded.

Normal probability plots indicated that both sample salaries are from normal populations. The summary statistics are given in the following table.

	Male	Female
sample size	23	19
mean	52.56	48.34 (in $1000)
variance	10.21	7.56 (in $1000)

a. Do these summary statistics indicate that the average salaries of the male programmers higher than that of female programmers?

b. Construct a 95% confidence interval for the mean difference between the salaries of male and female programmers.

Solution: Since the salaries are from normal populations, a two-sample procedure for comparing normal means is appropriate for this problem. Furthermore, to choose between the two comparison methods (one assumes that the population variances are equal and the other is not), we need to test the equality of the population variances. Using the dialog box [StatCalc→Continuous → Normal → Two-Sample Case → Test and CI for the Variance Ratio], we get the p-value for testing the equality of variances is 0.521731, which is greater than any practical level of significance. Therefore, the assumption that the variances are equal is tenable, and we can use the two-sample t procedures for the present problem.

a. Let μ_1 denote the mean salaries of all male programmers, and μ_2 denote the mean salaries of all female programmers in the company. We want to test

$$H_0 : \mu_1 \leq \mu_2 \text{ vs. } H_a : \mu_1 > \mu_2.$$

To compute the p-value for the above test, enter the sample sizes, means and standard deviations, click [p-values for] to get 2.58639e-005. Since this p-value is much less than any practical levels, we reject the null hypothesis, and conclude that the mean salaries of male programmers is higher than that of female programmers.

b. To compute a 95% confidence interval for $\mu_1 - \mu_2$, enter 0.95 for the confidence level, and click [2-sided] to get 2.33849 and 6.10151. That is, the mean difference is somewhere between $2338 and $6101.

10.5.3 Inference for the Difference between Two Means

Suppose that the test for equality of variances (Section 10.5.1) shows that the variances are significantly different, then the following Welch's approximate degrees of freedom method should be used to make inferences about $\mu_1 - \mu_2$. This approximate method is based on the result that

$$\frac{\bar{X}_1 - \bar{X}_2}{\sqrt{\frac{S_1^2}{n_1} + \frac{S_2^2}{n_2}}} \sim t_f \text{ approximately, with } f = \frac{\left(\frac{S_1^2}{n_1} + \frac{S_2^2}{n_2}\right)^2}{\left(\frac{S_1^4}{n_1^2(n_1-1)} + \frac{S_2^4}{n_2^2(n_2-1)}\right)}.$$

The hypothesis testing and interval estimation of $\mu_1 - \mu_2$ can be carried out as in Section 10.5.1 with the degrees of freedom f given above. For example, a $1 - \alpha$ confidence interval for $\mu_1 - \mu_2$ is given by

$$\bar{X}_1 - \bar{X}_2 \pm t_{f,1-\alpha/2}\sqrt{\frac{S_1^2}{n_1} + \frac{S_2^2}{n_2}},$$

where $t_{m,p}$ denotes the pth quantile of a t distribution with degrees of freedom f. This approximate method is commonly used, and the results based on this method are very accurate even for small samples.

The dialog box [StatCalc→Continuous → Normal → 2-sample → Test and CI for u1-u2] uses the above approximate method for hypothesis testing and interval estimation of $\mu_1 - \mu_2$ when the variances are not equal. Specifically, this dialog box computes confidence intervals and p-values for hypothesis testing about the difference between two normal means when the population variances are unknown and arbitrary.

Example 10.5.7 A sample of 8 observations from a normal population with mean μ_1, and variance σ_2^2 produced $\bar{X}_1 = 4$ and standard deviation $S_1 = 2.1$. A sample of 11 observations from another normal population with mean μ_2 and variance σ_2^2 yielded $\bar{X}_2 = 2$ with standard deviation $S_2 = 5$. The test for equality of variances [StatCalc→Continuous → Normal → Test and CI for the Variance Ratio] yielded a p-value of 0.0310413, and, hence, the assumption of equality of population variances seems to be invalid. Therefore, we should use the approximate degrees of freedom method described above.

To find a 95% confidence interval for $\mu_1 - \mu_2$, select [StatCalc→Continuous → Normal → 2-sample → Test and CI for u1-u2], enter the sample statistics and click [2-sided] to get $(-1.5985, 5.5985)$. To get one-sided limits, click [1-sided]. For this example, 95% one-sided lower limit is -0.956273, and 95% one-sided upper limit is 4.95627.

Suppose we want to test $H_0 : \mu_1 - \mu_2 = 0$ vs. $H_0 : \mu_1 - \mu_2 \neq 0$. To compute the p-values using *StatCalc*, click [p-values for] to get 0.253451. Thus, we can not conclude that the means are significantly different.

10.6 Tolerance Intervals

Let $X_1, \ldots, X_n$ be a sample from a normal population with mean μ and variance σ^2. Let $\bar{X}$ denote the sample mean and S denote the sample standard deviation.

10.6.1 Two-Sided Tolerance Intervals

For a given $0 < \beta < 1$, $0 < \gamma < 1$ and n, the tolerance factor k is to be determined so that the interval

$$\bar{X} \pm kS$$

would contain at least proportion β of the normal population with confidence γ. Mathematically, k should be determined so that

$$P_{\bar{X},S}\left\{P_X\left[X \in (\bar{X} - kS,\ \bar{X} + kS\ )|\bar{X}, S\right] \geq \beta\right\} = \gamma, \tag{10.6.1}$$

where X also follows the $N(\mu, \sigma^2)$ distribution independently of the sample. An explicit expression for k is not available and has to be computed numerically.

An approximate expression for k is given by

$$k \simeq \left(\frac{m\chi^2_{1,\beta}(1/n)}{\chi^2_{m,1-\gamma}}\right)^{1/2}, \tag{10.6.2}$$

where $\chi^2_{1,p}(1/n)$ denotes the pth quantile of a noncentral chi-square distribution with df $= 1$ and noncentrality parameter $1/n$, $\chi^2_{m,1-\gamma}$ denotes the $(1-\gamma)$th quantile of a central chi-square distribution with df $= m = n - 1$, the df associated with the sample variance. This approximation is extremely satisfactory even for small samples (as small as 3) if β and γ are greater than or equal to 0.95. [Wald and Wolfowitz 1946].

The dialog box [StatCalc→Continuous → Normal→ Tolerance Limits] uses an exact method [see Kendall and Stuart 1973, p. 134] of computing the tolerance factor k.

Remark 10.6.1 The k satisfies (10.6.1) is called the *tolerance factor*, and $\bar{X} \pm kS$ is called a (β, γ) *tolerance interval* or a β content – γ coverage tolerance interval.

10.6.2 One-Sided Tolerance Limits

The one-sided upper tolerance limit is given by $\bar{X} + cS$, where the tolerance factor c is to be determined so that

$$P_{\bar{X},S}\{P_X[X \le \bar{X} + cS|\bar{X}, S] \ge \beta\} = \gamma.$$

In this case, c is given by

$$c = \frac{1}{\sqrt{n}} t_{n-1,\gamma}(z_\beta \sqrt{n}),$$

where $t_{m,p}(\delta)$ denotes the 100pth percentile of a noncentral t distribution with df $= m$ and noncentrality parameter δ, and z_p denotes the 100pth percentile of the standard normal distribution. The same c can be used to find the lower tolerance limits; that is, if $\bar{X} + cS$ is the one-sided upper tolerance limits, then $\bar{X} - cS$ is the one-sided lower tolerance limit. The one-sided tolerance limits have interpretation similar to that of the two-sided tolerance limits. That is, at least 100β% of the data from the normal population are less than or equal to $\bar{X} + cS$ with confidence γ. Also, at least 100β% of the data are greater than or equal to $\bar{X} - cS$ with confidence γ.

Remark 10.6.2 The degrees of freedom associated with S^2 is $n - 1$. In some situations, the df associated with the S could be different from $n - 1$. For example, in one-way ANOVA, the df associated with the pooled sample variance S_p^2 is (total sample size $-$ g), where g denotes the number of groups. If one is interested in computing (β, γ) tolerance interval

$$\bar{X}_1 \pm k_1 S_p$$

for the first population, then for this case, the sample size is n_1 and the degrees of freedom associated with the pooled variance is

$$\sum_{i=1}^{g} n_i - g\ ,$$

where n_i denotes the size of the sample from the ith group, $i = 1, \ldots, g$.

For a given n, df, $0 < \beta < 1$ and $0 < \gamma < 1$, the dialog box [StatCalc→ Continuous → Normal → Tolerance Limits] computes the one-sided as well as two-sided tolerance factors.

Example 10.6.1 When $n = 23$, df $= 22$, $\beta = 0.90$, and $\gamma = 0.95$, the one-sided tolerance factor is 1.86902, and the two-sided tolerance factor is 2.25125. To compute the factors, enter 23 for [Sample Size n], 22 for [DF], 0.90 for [Proportion p] and 0.95 for [Coverage Prob g]; click [1-sided] to get 1.86902, and click [2-sided] to get 2.25125.

Applications

The normal-based tolerance factors are applicable to a non-normal distribution if it has a one–one relation with a normal distribution. For example, if X follows a lognormal distribution, then $\ln(X)$ follows a normal distribution. Therefore, the factors given in the preceding sections can be used to construct tolerance intervals for a lognormal distribution. Specifically, if the sample $Y_1, \ldots, Y_n$ is from a lognormal distribution, then normal based methods for constructing tolerance intervals can be used after taking logarithmic transformation of the sample. If U is a (β, γ) upper tolerance limit based on the logged data, then $\exp(U)$ is the (β, γ) upper tolerance limit for the lognormal distribution.

In many practical situations one wants to assess the proportion of the data fall in an interval or a region. For example, engineering products are usually required to satisfy certain tolerance specifications. The proportion of the products that are within the specifications can be assessed by constructing a suitable tolerance region based on a sample of products.

Example 10.6.2 Suppose that a lot of items submitted for inspection will be accepted if at least 95% of the items are within the specification (L, U), where L is the lower specification limit and U is the upper specification limit. In order to save time and cost, typically a sample of items is inspected and a (.95, .95) tolerance interval is constructed. If this tolerance interval falls in (L, U), then it can be concluded that at least 95% of the items in the lot are within the specification limits with 95% confidence, and, hence, the lot will be accepted.

In some situations, each item in the lot is required to satisfy only the lower specification. In this case, a (.95, .95) lower tolerance limit is constructed and compared with the lower specification L. If the lower tolerance limit is greater than or equal to L, then the lot will be accepted.

Example 10.6.3 Tolerance limits can be used to monitor exposure levels of employees to workplace contaminants. Specifically, if the upper tolerance limit based on exposure measurements from a sample of employees is less than a permissible exposure limit (PEL), then it indicates that a majority of the exposure measurements are within the PEL, and hence exposure monitoring might be reduced or terminated until a process change occurs. Such studies are feasible because the National Institute for Occupational Safety and Health provides PEL for many workplace chemicals. [Tuggle 1982].

Example 10.6.4 Let us construct tolerance limits for the exposure data given in Example 10.1.1. We already showed that the data satisfy the normality assumption. Note that the sample size is 15 (df = 15 − 1 = 14), the sample mean is 78.5 and the standard deviation is 15.9. The tolerance factor for a (.95, .95) tolerance interval is 2.96494. Using these numbers, we compute the tolerance interval as

$$78.5 \pm 2.96494 \times 15.9 = (31.4,\ 125.6).$$

That is, at least 95% of the exposure measurements fall between 31.4 and 125.6 with 95% confidence. The tolerance factor for a (.95, .95) one-sided limit is 2.566. The one-sided upper limit is $78.5 + 2.566 \times 15.9 = 119.3$. That is, at least 95% of the exposure measurements are below 119.3 with 95% confidence. The one-sided lower tolerance limit is $78.5 - 2.566 \times 15.9 = 37.7$. That is, at least 95% of the exposure measurements are above 37.7.

10.6.3 Equal-Tail Tolerance Intervals

Let $X_1, \ldots, X_n$ be a sample from a normal population with mean μ and variance σ^2. Let $\bar{X}$ denote the sample mean and S denote the sample standard deviation. The β content – γ coverage equal-tail tolerance interval (L, U) is constructed so that it would contain at least $100\beta\%$ of the "center data" of the normal population. That is, (L, U) is constructed such that not more than $100(1-\beta)/2\%$ of the data are less than L, and not more that $100(1-\beta)/2\%$ of the data are greater than U with confidence γ. This amounts to constructing (L, U) so that it would contain $\left(\mu - z_{\frac{1+\beta}{2}}\sigma,\ \mu + z_{\frac{1+\beta}{2}}\sigma\right)$ with confidence γ. Toward this, we consider the intervals of the form $(\bar{X} - kS,\ \bar{X} + kS)$, where k is to be determined so that

$$P\left(\bar{X} - kS < \mu - z_{\frac{1+\beta}{2}}\sigma \text{ and } \mu + z_{\frac{1+\beta}{2}}\sigma < \bar{X} + kS\right) = \gamma.$$

The dialog box [StatCalc→Continuous→Normal→Tol. Eq. Tails] uses an exact method due to Owen (1964) for computing the tolerance factor k satisfying the above probability requirement.

Example 10.6.5 In order to understand the difference between the tolerance interval and the equal-tail tolerance interval, let us consider Example 10.6.4 where we constructed (.95, .95) tolerance interval as (31.4, 125.6). Note that this interval would contain at least 95% of the data (not necessarily center data) with 95% confidence. Also, for this example, the sample size is 15, the sample mean is 78.5 and the standard deviation is 15.9. To compute the (.95, .95) equal-tail tolerance factor, enter 15 for [Sample Size n], 0.95 for [Proportion p] and 0.95 for [Coverage Prob g]; click [Tol Factor] to get 3.216. To find the tolerance interval, enter 78.5 for [x-bar], 15.9 for [s] and click [2-sided] to get (27.37, 129.6), which is $(\bar{X} - kS,\ \bar{X} + kS)$. We also observe that this equal-tai tolerance interval is wider than the tolerance interval (31.4, 125.6).

10.6.4 Simultaneous Hypothesis Testing for Quantiles

Let $X_1, \ldots, X_n$ be a sample from a normal population with mean μ and variance σ^2. Let $\bar{X}$ denote the sample mean and S denote the sample standard deviation Owen (1964) pointed out an acceptance sampling plan where a lot of items wil be accepted if the sample provides evidence in favor of the alternative hypothesis given below:

$$H_0: H_a^c \text{ vs. } H_a: L < \mu - z_{\frac{1+\beta}{2}}\sigma \text{ and } \mu + z_{\frac{1+\beta}{2}}\sigma < U,$$

where L and U are specified numbers, and β is a number in (0, 1), usually close to 1. Note that the lot is not acceptable if either

$$\mu - z_{\frac{1+\beta}{2}}\sigma \leq L,\ U \leq \mu - z_{\frac{1+\beta}{2}}\sigma \quad \text{or} \quad \mu - z_{\frac{1+\beta}{2}}\sigma \leq L \quad \text{and} \quad U \leq \mu - z_{\frac{1+\beta}{2}}\sigma.$$

The null hypothesis will be rejected at level α, if

$$L < \bar{X} - kS \quad \text{and} \quad \bar{X} + kS < U,$$

where k is to be determined such that

$$P(L < \bar{X} - kS \text{ and } \bar{X} + kS < U | H_0) = \alpha.$$

Notice that the factor k is determined in such a way that the probability of accepting an unacceptable lot (rejecting H_0 when it is true) is not more than α.

The dialog box [StatCalc→Continuous → Normal → Siml. Test for Quantiles] uses an exact method due to Owen (1964) for computing the factor k satisfying the above probability requirement.

Example 10.6.6 Let us use the summary statistics in Example 10.6.4 for illustrating above quantile test. Note that $n = 15$, $\bar{X} = 78.5$ and $S = 15.9$. We would like to test if the lower 2.5th percentile is greater than 30 and the upper 2.5th percentile is less than 128 at the level of 0.05. That is, our

$$H_0: H_a^c \text{ vs. } H_a: 30 < \mu - z_{.975}\sigma \text{ and } \mu + z_{.975}\sigma < 128.$$

Note that $(1 + \beta)/2 = 0.025$ implies $\beta = 0.95$. To find the factor k, enter 15 for the sample size, 0.95 for the proportion p, 0.05 for the level, and click [Tol

Factor] to get 2.61584. To get the limits, click on [2-sided] to get 36.9082. That is, $\bar{X} - kS = 36.9082$ and $\bar{X} + kS = 120.092$. Thus, we have enough evidence to conclude that the lower 2.5th percentile of the normal distribution is greater than 30 and the upper 2.5th percentile of the normal distribution is less than 128.

10.6.5 Tolerance Limits for One-Way Random Effects Model

The one-way random effects model is given by

$$X_{ij} = \mu + \tau_i + \varepsilon_{ij},\ j = 1{,}2{,}\ldots{,}n_i,\quad i = 1, 2, \ldots, k,$$

where X_{ij} is the jth measurement on the ith individual, μ is the overall fixed mean effect, the random effect $\tau_i \sim N(0, \sigma_\tau^2)$ independently of the error $\epsilon_{ij} \sim N(0, \sigma_e^2)$. Notice that $X_{ij} \sim N(\mu, \sigma_e^2 + \sigma_\tau^2)$.

A (β, γ) one-sided upper tolerance limit for the distribution of X_{ij} is the $100\gamma\%$ upper confidence limit for

$$\mu + z_\beta \sqrt{\sigma_e^2 + \sigma_\tau^2},$$

where z_β is the βth quantile of the standard normal distribution. Similarly, a (β, γ) one-sided lower tolerance limit for the distribution of X_{ij} is the $100\gamma\%$ lower confidence limit for

$$\mu - z_\beta \sqrt{\sigma_e^2 + \sigma_\tau^2}.$$

Define

$$N = \sum_{i=1}^{k} n_i,\ \bar{X}_i = \frac{1}{n_i} \sum_{j=1}^{n_i} X_{ij},\ \bar{\bar{X}} = \frac{1}{k} \sum_{i=1}^{k} \bar{X}_i \ \text{ and } \ \tilde{n} = \frac{1}{k} \sum_{i=1}^{k} \frac{1}{n_i}.$$

Furthermore, let

$$SS_e = \sum_{i=1}^{k} \sum_{j=1}^{n_i} (X_{ij} - \bar{X}_i)^2 \text{ and } SS_{\bar{x}} = \sum_{i=1}^{k} \left(\bar{X}_i - \bar{\bar{X}}\right)^2.$$

Let $t_{m,p}(c)$ denote the pth quantile of a noncentral t distribution with noncentrality parameter c and df $= m$. A (β, γ) upper tolerance limit is given by

$$\bar{\bar{X}} + t_{k-1,\gamma}(\delta) \sqrt{\frac{SS_{\bar{x}}}{k(k-1)}}, \tag{10.6.3}$$

where

$$\delta = z_\beta \left(k + \frac{k(k-1)(1-\tilde{n})}{N-k} \frac{SS_e}{SS_{\bar{x}}} F_{k-1,N-k,1-\gamma} \right)^{\frac{1}{2}},$$

and $F_{a,b,c}$ denotes the 100cth percentile of an F distribution with dfs a and b. The approximate tolerance limit in (10.6.3) is due to Krishnamoorthy and Mathew (2004), and is very accurate provided the F-statistic for the model is significant.

A (β, γ) lower tolerance limit is given by

$$\bar{\bar{X}} - t_{k-1,\gamma}(\delta)\sqrt{\frac{SS_{\bar{x}}}{k(k-1)}}.$$

For given summary statistics $\bar{\bar{X}}$, SS_e, $SS_{\bar{x}}$, $\tilde{n}$, N, k, and the values of the proportion β and coverage probability γ, the dialog box [StatCalc→Continuous →Normal→Tol Fac Rand Effects] computes the tolerance factors and one-sided tolerance limits.

Example 10.6.7 In this example, we consider $k = 5$ treatments. The data are given in the following table.

X_{1j}	X_{2j}	X_{3j}	X_{4j}	X_{5j}
2.977	3.525	4.191	5.834	0.73
0.914	4.094	0.398	5.370	4.011
2.666	3.797	3.206	2.596	2.033
-0.072	2.031	2.670	7.727	0.189
4.784	5.068	0.068	5.692	0.262
2.605	2.326	1.213	4.534	-3.700
	1.78	-0.385		0.685
	2.339			1.300
	4.306			-1.113
				2.734

The summary statistics are given in the following table.

n_i	$\bar{X}_i$	$(n_i - 1)S_i^2$
8	2.178	14.84
11	3.351	11.47
9	2.225	33.817
8	4.913	17.674
12	1.075	54.359
48	13.742	132.16

Other statistics are $\bar{\bar{X}} = 2.748$, $SS_e = 132.16$, $\tilde{n} = \left(\frac{1}{8} + \frac{1}{11} + \frac{1}{9} + \frac{1}{8} + \frac{1}{12}\right)/5 = 0.1071$ and $SS_{\bar{x}} = 8.449$. Also, note that $SS_e = \sum_{i=1}^{k} (n_i - 1)S_i^2$ and $N = 48$. To compute (0.90, 0.95) one-sided tolerance limits using *StatCalc*, enter 48 for N, 5 for k, 0.90 for proportion, 0.95 for coverage probability, 2.748 for [x-bar-bar], 132.16 for [SS_e], 8.449 for [SS_x], 0.1071 for [n~], and click on [1-sided] to get −2.64 and 8.14. This means that 90% of the data on X_{ij} exceed −2.64 with confidence 0.95; similarly, we can conclude that 90% of the data on X_{ij} are below 8.14 with confidence 0.95.

The above formulas and the method of constructing one-sided tolerance limits are also applicable for the *balanced* case (i.e., $n_1 = n_2 = \ldots = n_k$).

10.7 Properties and Results

1. Let $X_1, \ldots, X_n$ be independent normal random variables with $X_i \sim N(\mu_i, \sigma_i^2)$, $i = 1, 2, 3, \ldots, n$. Then

$$\sum_{i=1}^{n} a_i X_i \sim N\left(\sum_{i=1}^{n} a_i \mu_i, \sum_{i=1}^{n} a_i^2 \sigma_i^2\right),$$

   where $a_1, \ldots, a_n$ are constants.

2. Let U_1 and U_2 be independent uniform(0,1) random variables. Then

$$X_1 = \cos(2\pi U_1)\sqrt{-2\ln(U_2)},$$
$$X_2 = \sin(2\pi U_1)\sqrt{-2\ln(U_2)}$$

   are independent standard normal random variables [Box–Muller transformation].

3. Let Z be a standard normal random variable. Then, Z^2 is distributed as a chi-square random variable with df = 1.

4. Let X and Y be independent normal random variables with common variance but possibly different means. Define $U = X + Y$and $V = X - Y$. Then, U and V are independent normal random variables.

5. The sample mean

$$\bar{X} \text{ and } \{(X_1 - \bar{X}), \ldots, (X_n - \bar{X})\}$$

are statistically independent.

6. Let $X_1, \ldots, X_n$ be independent $N(\mu, \sigma^2)$ random variables. Then,

$$V^2 = \sum_{i=1}^{n} (X_i - \bar{X})^2 \text{ and } \left\{ \frac{(X_1 - \bar{X})}{V}, \ldots, \frac{(X_n - \bar{X})}{V} \right\}$$

are statistically independent.

7. Stein's (1981) Lemma: If X follows a normal distribution with mean μ and standard deviation σ, then

$$E(X - \mu)h(X) = \sigma E \left[\frac{\partial h(X)}{\partial X} \right]$$

provided the indicated expectations exist.

8. Let $X_1, \ldots, X_n$ be iid normal random variables. Then

$$\frac{\sum_{i=1}^{n} (X_i - \bar{X})^2}{\sigma^2} \sim \chi^2_{n-1} \quad \text{and} \quad \frac{\sum_{i=1}^{n} (X_i - \mu)^2}{\sigma^2} \sim \chi^2_n.$$

10.8 Relation to Other Distributions

1. Cauchy: If X and Y are independent standard normal random variables then $U = X/Y$ follows the Cauchy distribution (Chapter 26) with probability density function

$$f(u) = \frac{1}{\pi(1 + u^2)}, \quad -\infty < u < \infty.$$

2. F Distribution: If X and Y are independent standard normal random variables, then X^2 and Y^2 are independent and distributed as a χ^2 random variable with df $= 1$. Also $F = (X/Y)^2$ follows an $F_{1,1}$ distribution.

3. Gamma: Let Z be a standard normal random variable. Then

$$P(0 < Z \leq z) = P(Y < z^2)/2,$$

and

$$P(Y \leq y) = 2P(Z \leq \sqrt{y}) - 1, \quad y > 0,$$

where Y is a gamma random variable with shape parameter 0.5 and scale parameter 2.

4. Lognormal: A random variable Y is said to have a lognormal distribution with parameters μ and σ if $\ln(Y)$ follows a normal distribution. Therefore,

$$P(Y \leq x) = P(\ln(Y) \leq \ln(x)) = P(X \leq \ln(x)),$$

where X is the normal random variable with mean μ and standard deviation σ.

For more results and properties, see Patel and Read (1981).

10.9 Random Number Generation

Algorithm 10.9.1

Generate U_1 and U_2 from uniform(0,1) distribution. Set

$$X_1 = \cos(2\pi U_1)\sqrt{-2\ln(U_2)}$$
$$X_2 = \sin(2\pi U_1)\sqrt{-2\ln(U_2)}.$$

Then X_1 and X_2 are independent N(0, 1) random variables. There are several other methods available for generating normal random numbers (see Kennedy and Gentle 1980, Section 6.5). The above Box–Muller transformation is simple to implement and is satisfactory if it is used with a good uniform random number generator.

The following algorithm due to Kinderman and Ramage (1976; correction Vol. 85, p. 272) is faster than the Box–Muller transformations. For better accuracy, double precision may be required.

[1] *Algorithm 10.9.2*

[1] Reproduced with permission from the American Statistical Association.

In the following, u, v, and w are independent uniform(0, 1) random numbers.

The output x is a N(0, 1) random number.

Set g = 2.21603 58671 66471

$$f(t) = \frac{1}{\sqrt{2\pi}} \exp(-t^2/2) - 0.180025191068563(g - |t|), \quad |t| < g$$

```
      Generate u
      If u < 0.88407 04022 98758, generate v
      return x = g*(1.3113 16354 44180*u + v + 1)

      If u < 0.97331 09541 73898 go to 4

3     Generate v and w
      Set t = g**2/2 - ln(w)
      If v**2*t > g**2/2, go to 3

      If u < 0.98665 54770 86949, return x = sqrt(2*t)
      else return x = - sqrt(2*t)

4     If u < 0.95872 08247 90463 goto 6

5     Generate v and w
      Set z = v - w
          t = g - 0.63083 48019 21960*min(v, w)
      If max(v, w) <= 0.75559 15316 67601 goto 9
      If 0.03424 05037 50111*abs(z) <= f(t), goto 9
      goto 5

6     If u < 0.91131 27802 88703 goto 8

7      Generate v and w
       Set z = v - w
           t = 0.47972 74042 22441 + 1.10547 36610 22070*min(v, w)
       If max(v, w) <= 0.87283 49766 71790, goto 9
       If 0.04926 44963 73128*abs(z) <= f(t), goto 9
       goto 7

8      Generate v and w
```

```
Set z = v - w
    t = 0.47972 74042 22441 - 0.59950 71380 15940*min(v, w)
If max(v, w) <= 0.80557 79244 23817 goto 9
If t >= 0 and 0.05337 75495 06886*abs(z) <= f(t), goto 9
goto 8

9       If z < 0, return x = t
        else return x = -t
```

10.10 Computing the Distribution Function

For 0 < z < 7, the following polynomial approximation can be used to compute the probability $\Phi(z) = P(Z \le z)$.

$$\Phi(z) = e^{\frac{-z^2}{2}} \frac{P_7 z^7 + P_6 z^6 + P_5 z^5 + P_4 z^4 + P_3 z^3 + P_2 z^2 + P_1 z + P_0}{Q_8 z^8 + Q_7 z^7 + Q_6 z^6 + Q_5 z^5 + Q_4 z^4 + Q_3 z^3 + Q_2 z^2 + Q_1 z + Q_0},$$

where

$P_0 = 913.167442114755700$, $P_1 = 1024.60809538333800$,
$P_2 = 580.109897562908800$, $P_3 = 202.102090717023000$,
$P_4 = 46.0649519338751400$, $P_5 = 6.81311678753268400$,
$P_6 = 6.047379926867041E-1$, $P_7 = 2.493381293151434E-2$,
and
$Q_0 = 1826.33488422951125$, $Q_1 = 3506.420597749092$,
$Q_2 = 3044.77121163622200$, $Q_3 = 1566.104625828454$,
$Q_4 = 523.596091947383490$, $Q_5 = 116.9795245776655$,
$Q_6 = 17.1406995062577800$, $Q_7 = 1.515843318555982$,
$Q_8 = 6.25E-2$.

For $z \ge 7$, the following continued fraction can be used to compute the probabilities.

$$\Phi(z) = 1 - \varphi(z) \left[\frac{1}{z+} \frac{1}{z+} \frac{2}{z+} \frac{3}{z+} \frac{4}{z+} \frac{5}{z+} \cdots \right],$$

where φ(z) denotes the standard normal density function. The above method is supposed to give 14 decimal accurate probabilities. [Hart et al. 1968, p. 137].

The following Fortran function subroutine for evaluating the standard normal cdf is based on the above computational method.

```
      double precision function gaudf(x)
      implicit doubleprecision (a-h,o-z)
      dimension p(8), q(9)
      logical check
      data p /913.167442114755700d0, 1024.60809538333800d0,
     +       580.109897562908800d0, 202.102090717023000d0,
     +       46.0649519338751400d0, 6.81311678753268400d0,
     +       6.047379926867041d-1,2.493381293151434d-2/

      data q /1826.33488422951125d0, 3506.420597749092d0,
     +       3044.77121163622200d0, 1566.104625828454d0,
     +       523.596091947383490d0, 116.9795245776655d0,
     +       17.1406995062577800d0, 1.515843318555982d0,
     +       6.25d-2/

      sqr2pi = 2.506628274631001d0;

      z = max(x,-x)
      check = .false.
      if(x > 0.0d0) check = .true.

      prob = 0.0d0
      if (z > 32.0d0) goto 1

      first = dexp(-0.5d0*z*z)
      phi = first/sqr2pi

      if (z < 7.0d0) then
         prob = first*(((((((p(8)*z + p(7))*z + p(6))*z
     +   + p(5))*z + p(4))*z+ p(3))*z + p(2))*z + p(1))/
     +     ((((((((q(9)*z + q(8))*z + q(7))*z + q(6))*z
     +   + q(5))*z + q(4))*z + q(3))*z + q(2))*z + q(1))
      else
         prob = phi/(z + 1.0/(z + 2.0/(z + 3.0/(z + 4.0/
     +   (z + 5.0/(z + 6.0/(z + 7.0)))))))
      end if
1     if (check) prob = 1.0 - prob
      gaudf = prob

      end
```

Chapter 11

Chi-Square Distribution

11.1 Description

Let $X_1, \ldots, X_n$ be independent standard normal random variables. The distribution of

$$X = \sum_{i=1}^{n} X_i^2$$

is called the chi-square distribution with degrees of freedom (df) n, and its probability density function is given by

$$f(x|n) = \frac{1}{2^{n/2}\Gamma(n/2)} e^{-x/2} x^{n/2-1}, \qquad x > 0,\ n > 0. \tag{11.1.1}$$

The chi-square random variable with df $= n$ is denoted by χ_n^2. Since the probability density function is valid for any $n > 0$, alternatively, we can define the chi-square distribution as the one with the probability density function (11.1.1). This latter definition holds for any $n > 0$. An infinite series expression for the cdf is given in Section 11.5.1.

Plots in Figure 11.1 indicate that, for large degrees of freedom m, the chi-square distribution is symmetric about its mean. Furthermore, χ_a^2 is stochastically larger than χ_b^2 for $a > b$.

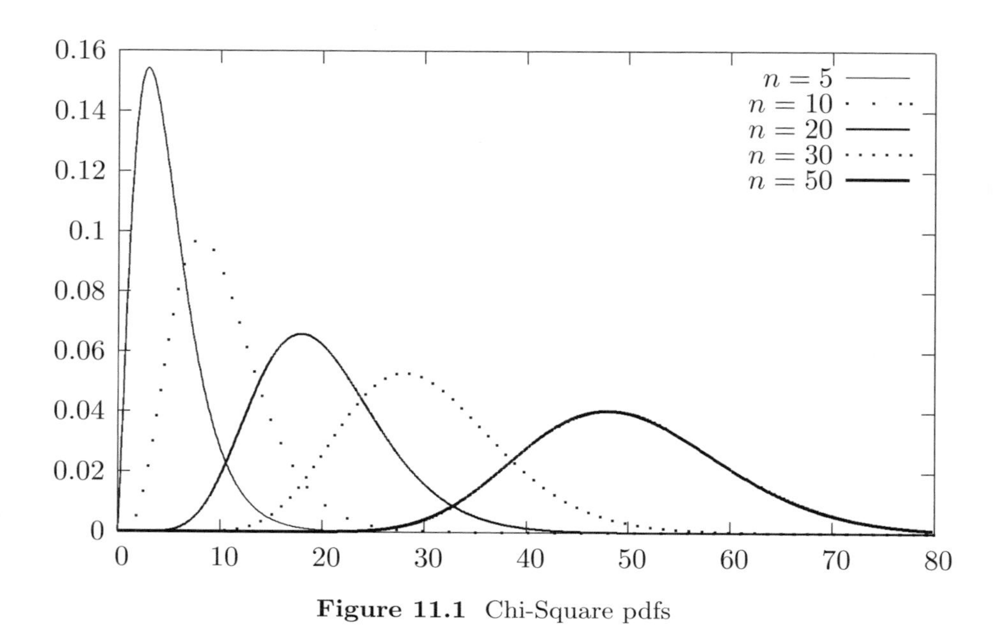

Figure 11.1 Chi-Square pdfs

11.2 Moments

Mean:	n
Variance:	$2n$
Mode:	$n-2, \quad n > 2.$
Coefficient of Variation:	$\sqrt{\frac{2}{n}}$
Coefficient of Skewness:	$2\sqrt{\frac{2}{n}}$
Coefficient of Kurtosis:	$3+\frac{12}{n}$
Mean Deviation:	$\frac{n^{n/2}e^{-n/2}}{2^{n/2-1}\Gamma(n/2)}$
Moment Generating Function:	$(1-2t)^{-n/2}$
Moments about the Origin:	$E[(\chi_n^2)^k] = 2^k \prod_{i=0}^{k-1}(n/2+i),$ $k = 1, 2, \cdots$

11.3 Computing Table Values

The dialog box [StatCalc→Continuous → Chi-sqr] computes the probabilities and percentiles of a chi-square distribution. For the degrees of freedom greater than 100,000, a normal approximation to the chi-square distribution is used to compute the cdf as well as the percentiles.

To compute probabilities: Enter the value of the degrees of freedom (df), and the value of x at which the cdf is to be computed; click P(X <= x).

Example 11.3.1 When df = 13.0 and $x = 12.3$,

$$P(X \leq 12.3) = 0.496789 \text{ and } P(X > 12.3) = 0.503211.$$

To compute percentiles: Enter the values of the degrees of freedom and the cumulative probability, and click [x].

Example 11.3.2 When df = 13.0 and the cumulative probability = 0.95, the 95th percentile is 22.362. That is, $P(X \leq 22.362) = 0.95$.

To compute the df: Enter the values of the cumulative probability and x, and click [DF].

Example 11.3.3 When $x = 6.0$ and the cumulative probability = 0.8, the value of DF is 4.00862.

To compute moments: Enter the value of the df and click [M].

11.4 Applications

The chi-square distribution is also called the variance distribution by some authors, because the variance of a random sample from a normal distribution follows a chi-square distribution. Specifically, if $X_1, \ldots, X_n$ is a random sample from a normal distribution with mean μ and variance σ^2, then

$$\frac{\sum\limits_{i=1}^{n} (X_i - \bar{X})^2}{\sigma^2} = \frac{(n-1)S^2}{\sigma^2} \sim \chi^2_{n-1}.$$

This distributional result is useful to make inferences about σ^2. (see Section 10.4).

In categorical data analysis consists of an $r \times c$ table, the usual test statistic,

$$T = \sum_{i=1}^{r} \sum_{j=1}^{c} \frac{(O_{ij} - E_{ij})^2}{E_{ij}} \sim \chi^2_{(r-1)\times(c-1)},$$

where O_{ij} and E_{ij} denote, respectively, the observed and expected cell frequencies. The null hypothesis of independent attributes will be rejected at a level of significance α, if an observed value of T is greater than $(1-\alpha)$th quantile of a chi-square distribution with df $= (r-1) \times (c-1)$.

The chi-square statistic

$$\sum_{i=1}^{k} \frac{(O_i - E_i)^2}{E_i}$$

can be used to test whether a frequency distribution fits a specific model. See Section 1.4.2 for more details.

11.5 Properties and Results

11.5.1 Properties

1. If $X_1, \ldots, X_k$ are independent chi-square random variables with degrees of freedom $n_1, \ldots, n_k$, respectively, then

$$\sum_{i=1}^{k} X_i \sim \chi^2_m \text{ with } m = \sum_{i=1}^{k} n_i.$$

2. Let Z be a standard normal random variable. Then $Z^2 \sim \chi^2_1$.

3. Let $F(x|n)$ denote the cdf of χ^2_n. Then

   a. $F(x|n) = \frac{1}{\Gamma(n/2)} \sum_{i=0}^{\infty} \frac{(-1)^i (x/2)^{n/2+i}}{i!\Gamma(n/2+i)}$,

   b. $F(x|n+2) = F(x|n) - \frac{(x/2)^{n/2} e^{-x/2}}{\Gamma(n/2+1)}$,

   c. $F(x|2n) = 1 - 2 \sum_{k=1}^{n} f(x|2k)$,

   d. $F(x|2n+1) = 2\Phi(\sqrt{x}) - 1 - 2 \sum_{k=1}^{n} f(x|2k+1)$,

   where $f(x|n)$ is the probability density function of χ^2_n, and Φ denotes the cdf of the standard normal random variable. [(a) Abramowitz and Stegun 1965, p. 941; (b) and (c) Peizer and Pratt 1968; (d) Puri 1973]

4. Let $\mathbf{Z}' = (Z_1, \ldots, Z_m)'$ be a random vector whose elements are independent standard normal random variables, and A be an $m \times m$ symmetric matrix with rank $= k$. Then

$$Q = Z'AZ = \sum_{i=1}^{m} \sum_{j=1}^{m} a_{ij} Z_i Z_j \sim \chi_k^2$$

if and only if A is an idempotent matrix, that is, $A^2 = A$.

5. Cochran's Theorem: Let $\mathbf{Z}$ be as defined in (4) and A_i be an $m \times m$ symmetric matrix with $\text{rank}(A_i) = k_i, \quad i = 1, 2, \ldots, r$. Let

$$Q_i = \mathbf{Z}' A_i \mathbf{Z}, \quad i = 1, 2, \ldots, r$$

and

$$\sum_{i=1}^{m} Z_i^2 = \sum_{i=1}^{r} Q_i.$$

Then $Q_1, \ldots, Q_r$ are independent with $Q_i \sim \chi_{k_i}^2, \quad i = 1, 2, \ldots, r$, if and only if

$$\sum_{i=1}^{r} k_i = m.$$

6. For any real valued function f,

$$E[(\chi_n^2)^k f(\chi_n^2)] = \frac{2^k \Gamma(n/2 + k)}{\Gamma(n/2)} E[f(\chi_{n+2k}^2)],$$

provided the indicated expectations exist.

7. Haff's (1979) Identity: Let f and h be real valued functions, and X be a chi-square random variable with df $= n$. Then

$$E[f(X)h(X)] = 2E\left[f(X)\frac{\partial h(X)}{\partial X}\right] + 2E\left[\frac{\partial f(X)}{\partial X}h(X)\right] + (n-2)E\left[\frac{f(X)h(X}{X}\right.$$

provided the indicated expectations exist.

11.5.2 Relation to Other Distributions

1. F and Beta: Let X and Y be independent chi-square random variables with degrees of freedoms m and n, respectively. Then

$$\frac{(X/m)}{(Y/n)} \sim F_{m,n}.$$

Furthermore, $\frac{X}{X+Y} \sim$ beta$(m/2, n/2)$ distribution.

2. Beta: If $X_1, \ldots, X_k$ are independent chi-square random variables with degrees of freedoms $n_1, \ldots, n_k$, respectively. Define

$$W_i = \frac{X_1 + \ldots + X_i}{X_1 + \ldots + X_{i+1}}, \quad i = 1, 2, \ldots, k-1.$$

The random variables $W_1, \ldots, W_{k-1}$ are independent with

$$W_i \sim \text{beta}\left(\frac{m_1 + \ldots + m_i}{2}, \frac{m_{i+1}}{2}\right), \quad i = 1, 2, \ldots, k-1.$$

3. Gamma: The gamma distribution with shape parameter a and scale parameter b specializes to the chi-square distribution with df $= n$ when $a = n/2$ and $b = 2$. That is, gamma$(n/2, 2) \sim \chi^2_n$.

4. Poisson: Let χ^2_n be a chi-square random variable with even degrees of freedom n. Then

$$P(\chi^2_n > x) = \sum_{k=0}^{n/2-1} \frac{e^{-x/2}(x/2)^k}{k!}.$$

[see Section 15.1]

5. t distribution: See Section 13.4.1.

6. Laplace: See Section 20.6.

7. Uniform: See Section 9.4.

11.5.3 Approximations

1. Let Z denote the standard normal random variable.

   a. $P(\chi^2_n \le x) \simeq P(Z \le \sqrt{2x} - \sqrt{2n-1}), \ \ n > 30.$

   b. $P(\chi^2_n \le x) \simeq P\left(Z \le \sqrt{\frac{9n}{2}}\left[\left(\frac{x}{n}\right)^{1/3} - 1 + \frac{2}{9n}\right]\right).$

   c. Let X denote the chi-square random variable with df $= n$. Then

$$\frac{X - n + 2/3 - 0.08/n}{|X - n + 1|}\left((n-1)\ln\left(\frac{n-1}{X}\right) + X - n + 1\right)^{1/2}$$

   is approximately distributed as a standard normal random variable.

[Peizer and Pratt 1968]

2. Let $\chi^2_{n,p}$ denote the pth percentile of a χ^2_n distribution, and z_p denote the pth percentile of the standard normal distribution. Then

a. $\chi^2_{n,p} \simeq \frac{1}{2}\left(z_p + \sqrt{2n-1}\right)^2, \quad n > 30.$

b. $\chi^2_{n,p} \simeq n\left(1 - \frac{2}{9n} + z_p\sqrt{\frac{2}{9n}}\right)^3.$

The approximation (b) is satisfactory even for small n. [Wilson and Hilferty 1931]

11.6 Random Number Generation

For smaller degrees of freedom, the following algorithm is reasonably efficient.

Algorithm 11.6.1

Generate $U_1, \ldots, U_n$ from uniform(0, 1) distribution.
Set $X = -2(\ln U_1 + \ldots + \ln U_n)$.

Then, X is a chi-square random number with df $= 2n$. To generate chi-square random numbers with odd df, add one Z^2 to X, where $Z \sim N(0,1)$. (see Section 11.5.1)

Since the chi-square distribution is a special case of the gamma distribution with the shape parameter $a = n/2$, and the scale parameter $b = 2$, the algorithms for generating gamma variates can be used to generate the chi-square variates (see Section 15.7).

11.7 Computing the Distribution Function

The distribution function and the percentiles of the chi-square random variable can be evaluated as a special case of the gamma($n/2$, 2) distribution (see Section 15.8). Specifically,

$$P(\chi^2_n \leq x|n) = P(Y \leq x|n/2, 2),$$

where Y is a gamma$(n/2, 2)$ random variable.

Chapter 12

F Distribution

12.1 Description

Let X and Y be independent chi-square random variables with degrees of freedoms m and n, respectively. The distribution of the ratio

$$F_{m,n} = \frac{\left(\frac{X}{m}\right)}{\left(\frac{Y}{n}\right)}$$

is called the F distribution with the numerator df $= m$ and the denominator df $= n$. The probability density function of an $F_{m,n}$ distribution is given by

$$f(x|m,n) = \frac{\Gamma\left(\frac{m+n}{2}\right)}{\Gamma\left(\frac{m}{2}\right)\Gamma\left(\frac{n}{2}\right)} \frac{\left(\frac{m}{2}\right)^{m/2} x^{m/2-1}}{\left(\frac{n}{2}\right)^{m/2}\left[1+\frac{mx}{n}\right]^{m/2+n/2}}, \quad m > 0, n > 0, x > 0.$$

Let S_i^2 denote the variance of a random sample of size n_i from a $N(\mu_i, \sigma^2)$ distribution, $i = 1, 2$. Then the variance ratio S_1^2/S_2^2 follows an F_{n_1-1,n_2-1} distribution. For this reason, the F distribution is also known as the variance ratio distribution.

We observe from the plots of pdfs in Figure 12.1 that the F distribution is always skewed to right; also, for equally large values of m and n, the F distribution is approximately symmetric about unity.

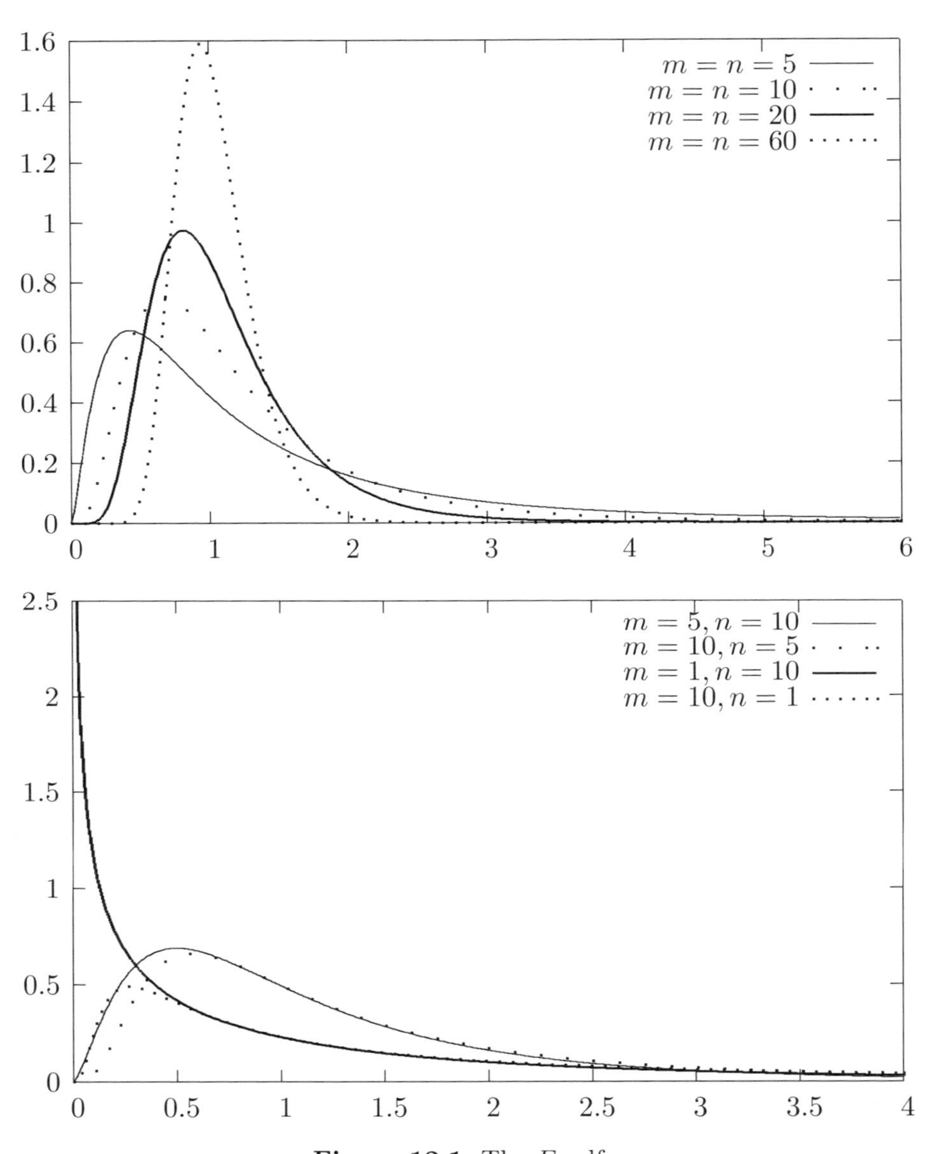

Figure 12.1 The F pdfs

12.2 Moments

Mean:	$\frac{n}{n-2}$
Variance:	$\frac{2n^2(m+n-2)}{m(n-2)^2(n-4)}, \quad n > 4.$
Mode:	$\frac{n(m-2)}{m(n+2)}, \quad m > 2.$
Moment Generating Function:	does not exist.
Coefficient of Variation:	$\frac{\sqrt{2(m+n-2)}}{\sqrt{m(n-4)}}, \quad n > 4.$
Coefficient of Skewness:	$\frac{(2m+n-2)\sqrt{8(n-4)}}{(n-6)\sqrt{m(m+n-2)}}, \quad n > 6.$
Coefficient of Kurtosis:	$3 + \frac{12[(n-2)^2(n-4)+m(m+n-2)(5n-22)]}{m(n-6)(n-8)(m+n-2)}, \quad n > 8.$
Moments about the Origin:	$\frac{\Gamma(m/2+k)\Gamma(n/2-k)}{\Gamma(m/2)\Gamma(n/2)}(n/m)^k,$ $n > 2k, \quad k = 1, 2, ...$

12.3 Computing Table Values

The dialog box [StatCalc→Continuous→F] computes probabilities, percentiles, moments and also the degrees of freedoms when other parameters are given.

To compute probabilities: Enter the numerator df, denominator df, and the value x at which the cdf is to be evaluated; click [P(X <= x)].

Example 12.3.1 When the numerator df = 3.3, denominator df = 44.5 and the observed value $x = 2.3$, $P(X \leq 2.3) = 0.915262$ and $P(X > 2.3) = 0.084738$.

To compute percentiles: Enter the values of the degrees of freedoms and the cumulative probability; click [x].

Example 12.3.2 When the numerator df = 3.3, denominator df = 44.5 and the cumulative probability = 0.95, the 95th percentile is 2.73281. That is, $P(X \leq 2.73281) = 0.95$.

To compute other parameters: *StatCalc* also computes the df when other values are given.

Example 12.3.3 When the numerator df = 3.3, cumulative probability = 0.90, $x = 2.3$ and the value of the denominator df = 22.4465. To find this value, enter other known values in appropriate edit boxes, and click on [Den DF].

To compute moments: Enter the values of the numerator df, denominator df, and click [M].

12.4 Properties and Results

12.4.1 Identities

1. For $x > 0$, $P(F_{m,n} \leq x) = P(F_{n,m} \geq 1/x)$.

2. If $F_{m,n,p}$ is the pth quantile of an $F_{m,n}$ distribution, then

$$F_{n,m,1-p} = \frac{1}{F_{m,n,p}}.$$

12.4.2 Relation to Other Distributions

1. Binomial: Let X be a binomial(n, p) random variable. For a given k

$$P(X \geq k|n,p) = P\left(F_{2k,2(n-k+1)} \leq \frac{(n-k+1)p}{k(1-p)}\right).$$

2. Beta: Let $X = F_{m,n}$. Then

$$\frac{mX}{n + mX}$$

follows a beta$(m/2, n/2)$ distribution.

3. Student's t: $F_{1,n}$ is distributed as t_n^2, where t_n denotes Student's t variable with df = n.

4. Laplace: See Section 20.6.

12.4.3 Series Expansions

For $y > 0$, let $x = \frac{n}{n+my}$.

1. For even m and any positive integer n,

$$\begin{aligned} P(F_{m,n} \leq y) &= 1 - x^{(m+n-2)/2} \left\{ 1 + \frac{m+n-2}{2} \left(\frac{1-x}{x} \right) \right. \\ &+ \frac{(m+n-2)(m+n-4)}{2 \cdot 4} \left(\frac{1-x}{x} \right)^2 \\ &+ \left. \frac{(m+n-2) \cdots (n+2)}{2 \cdot 4 \cdot \cdots \cdot (m-2)} \left(\frac{1-x}{x} \right)^{(m-2)/2} \right\}. \end{aligned}$$

2. For even n and any positive integer m,

$$\begin{aligned} P(F_{m,n} \leq y) &= (1-x)^{(m+n-2)/2} \left\{ 1 + \frac{m+n-2}{2} \left(\frac{x}{1-x} \right) \right. \\ &+ \frac{(m+n-2)(m+n-4)}{2 \cdot 4} \left(\frac{x}{1-x} \right)^2 + \ldots \\ &+ \left. \frac{(m+n-2) \cdots (m+2)}{2 \cdot 4 \cdot \cdots \cdot (n-2)} \left(\frac{x}{1-x} \right)^{(n-2)/2} \right\}. \end{aligned}$$

3. Let $\theta = \arctan\left(\sqrt{\frac{my}{n}}\right)$. For odd n,

   (a) $P(F_{1,1} \leq y) = \frac{2\theta}{\pi}$.

   (b) $P(F_{1,n} \leq y) = \frac{2}{\pi}\left\{\theta + \sin(\theta)\left[\cos(\theta) + \frac{2}{3}\cos^3(\theta) + \ldots + \frac{2 \cdot 4 \cdots (n-3)}{3 \cdot 5 \cdots (n-2)}\cos^{n-2}(\theta)\right]\right\}$

   (c) For odd m and any positive integer n,

$$\begin{aligned} P(F_{m,n} \leq y) &= \frac{2}{\pi} \left\{ \theta + \sin(\theta) \left[\cos(\theta) + \frac{2\cos^3(\theta)}{3} + \cdots \right. \right. \\ &+ \left. \left. \frac{2 \cdot 4 \cdots (n-3)}{3 \cdot 5 \cdots (n-2)} \cos^{n-2}(\theta) \right] \right\} \\ &- \frac{2[(n-1)/2]!}{\sqrt{\pi}\Gamma(n/2)} \sin(\theta)\cos^n(\theta) \times \left\{ 1 + \frac{n+1}{3} \sin^2(\theta) + \cdots \right. \\ &+ \left. \frac{(n+1)(n+3) \cdots (m+n-4)}{3 \cdot 5 \cdots (m-2)} \sin^{m-3}(\theta) \right\}. \end{aligned}$$

[Abramowitz and Stegun 1965, p. 946]

12.4.4 Approximations

1. For large m, $\frac{n}{F_{m,n}}$ is distributed as χ^2_n.

2. For large n, $mF_{m,n}$ is distributed as χ^2_m.

3. Let $M = n/(n-2)$. For large m and n,

$$\frac{F_{n,m} - M}{M\sqrt{\frac{2(m+n-2)}{m(n-4)}}}$$

   is distributed as the standard normal random variable. This approximation is satisfactory only when both degrees of freedoms are greater than or equal to 100.

4. The distribution of

$$Z = \frac{\sqrt{(2n-1)mF_{m,n}/n} - \sqrt{2m-1}}{\sqrt{1 + mF_{m,n}/n}}$$

   is approximately standard normal. This approximation is satisfactory even for small degrees of freedoms.

5. $\frac{F^{1/3}\left(1-\frac{2}{9n}\right)-\left(1-\frac{2}{9m}\right)}{\sqrt{\frac{2}{9m}+F^{2/3}\frac{2}{9n}}} \sim N(0,1)$ approximately.

[Abramowitz and Stegun 1965, p. 947]

12.5 Random Number Generation

Algorithm 12.5.1

For a given m and n:
Generate X from gamma$(m/2, 2)$ (see Section 15.7)
Generate Y from gamma$(n/2, 2)$
Set $F = nX/(mY)$.

F is the desired random number from the F distribution with numerator df $= m$, and the denominator df $= n$.

Algorithm 12.5.2

Generate Y from a beta($m/2$, $n/2$) distribution (see Section 16.7), and set

$$F = \frac{nY}{m(1-Y)}.$$

F is the desired random number from the F distribution with numerator df $= m$, and the denominator df $= n$.

12.6 A Computational Method for Probabilities

For smaller degrees of freedoms, the distribution function of $F_{m,n}$ random variable can be evaluated using the series expansions given in Section 12.4. For other degrees of freedoms, algorithm for evaluating the beta distribution can be used. Probabilities can be computed using the relation that

$$P(F_{m,n} \le x) = P\left(Y \le \frac{mx}{n+mx}\right),$$

where Y is the beta$(m/2, n/2)$ random variable. The pth quantile of an $F_{m,n}$ distribution can be computed using the relation that

$$F_{m,n,p} = \frac{n\,\text{beta}^{-1}(p; m/2, n/2)}{m(1-\text{beta}^{-1}(p; m/2, n/2))},$$

where beta$^{-1}(p; a, b)$ denotes the pth quantile of a beta(a, b) distribution.

Chapter 13

Student's t Distribution

13.1 Description

Let Z and S be independent random variables such that

$$Z \sim N(0,1) \quad \text{and} \quad nS^2 \sim \chi_n^2.$$

The distribution of $t = Z/S$ is called Student's t distribution with df $= n$. The Student's t random variable with df $= n$ is commonly denoted by t_n, and its probability density function is

$$f(x|n) = \frac{\Gamma[(n+1)/2]}{\Gamma(n/2)\sqrt{n\pi}} \frac{1}{(1+x^2/n)^{(n+1)/2}}, \quad -\infty < x < \infty,\ n \geq 1.$$

Probability density plots of t_n are given in Figure 13.1 for various degrees of freedoms. We observe from the plots that for large n, t_n is distributed as the standard normal random variable.

Series expansions for computing the cdf of t_n are given in Section 13.5.3.

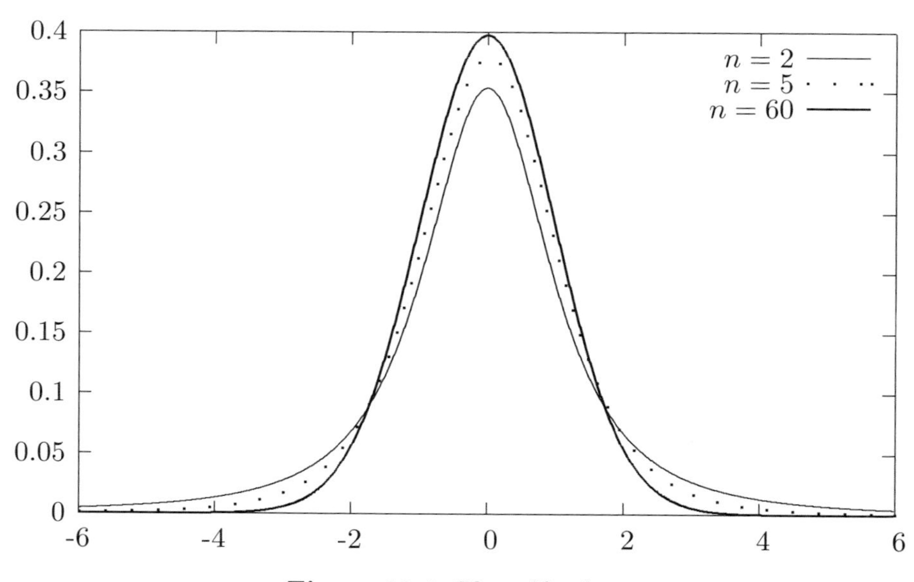

Figure 13.1 The pdfs of t_n

13.2 Moments

Mean:	0 for $n > 1$; undefined for $n = 1$.
Variance:	$n/(n-2), \quad n > 2.$
Median:	0
Mode:	0
Mean Deviation:	$\frac{\sqrt{n}\,\Gamma((n-1)/2)}{\sqrt{\pi}\,\Gamma(n/2)}$
Coefficient of Skewness:	0
Coefficient of Kurtosis:	$\frac{3(n-2)}{(n-4)}, \quad n > 4.$
Moment Generating Function:	does not exist
Moments about the Origin:	$E(t_n^k) = \begin{cases} 0 \text{ for odd } k < n, \\ \frac{1\cdot 3\cdot 5\cdots(k-1)}{(n-2)(n-4)\ldots(n-k)} n^{k/2} \\ \text{for even } k < n. \end{cases}$

13.3 Computing Table Values

The dialog box [StatCalc→Continuous→ Student t] computes probabilities, percentiles, moments and also the degrees of freedom for given other values.

To compute probabilities: Enter the value of the degrees of freedom (df), and the observed value x; click [x].

Example 13.3.1 When df = 12.0 and the observed value $x = 1.3$,

$$P(X \leq 1.3) = 0.890991 \text{ and } P(X > 1.3) = 0.109009.$$

To compute percentiles: Enter the value of the degrees of freedom, and the cumulative probability; click [x].

Example 13.3.2 When df = 12.0, and the cumulative probability = 0.95, the 95th percentile is 1.78229. That is, $P(X \leq 1.78229) = 0.95$.

To compute the DF: Enter the value of x, and the cumulative probability; click [DF].

Example 13.3.3 When $x = 1.3$, and the cumulative probability = 0.9, the value of DF = 46.5601.

To compute moments: Enter the value of the df and click [M].

13.4 Distribution of the Maximum of Several $|t|$ Variables

Let $X_1, \ldots, X_k$ be independent normal random variables with mean μ and common standard deviation σ. Let mS^2/σ^2 follow a chi-square distribution with df = m. The dialog box [StatCalc→Continuous→Student's t→ Max $|t|$] computes the distribution function of

$$X = \max_{1 \leq i \leq k} \left\{ \frac{|X_i|}{S} \right\} = \max_{1 \leq i \leq k} \{|t_i|\}, \tag{13.4.1}$$

where $t_1, ..., t_k$ are Student's t variables with df = m. The percentiles of X are useful for constructing simultaneous confidence intervals for the treatment effects and orthogonal estimates in the analysis of variance, and to test extreme values.

13.4.1 An Application

One-Way Analysis of Variance

Suppose we want to compare the effects of k treatments in a one-way analysis of variance setup based on the following summary statistics:

treatments	1	...	k
sample sizes	n_1	...	n_k
sample means	$\bar{X}_1$	...	$\bar{X}_k$
sample variances	S_1^2	...	S_k^2

Let $n = \sum_{i=1}^{k} n_i$, and $S_p^2 = \sum_{i=1}^{k} \frac{(n_i-1)S_i^2}{n-k}$ be the pooled sample variance, and

$$\bar{\bar{X}} = \frac{\sum_{i=1}^{k} n_i \bar{X}_i}{n}$$

be the pooled sample mean.

For testing $H_0 : \mu_1 = ... = \mu_k$ vs. $H_a : \mu_i \neq \mu_j$ for some $i \neq j$, the F statistic is given by

$$\frac{\sum_{i=1}^{k} n_i(\bar{X}_i - \bar{\bar{X}})^2/(k-1)}{S_p^2},$$

which follows an F distribution with numerator df $= k-1$ and the denominator df $= n-k$. For an observed value F_0 of the F statistic, the null hypothesis will be rejected if $F_0 > F_{k-1,n-k,1-\alpha}$, where $F_{k-1,n-k,1-\alpha}$ denotes the $(1-\alpha)$th quantile of an F distribution with the numerator df $= k-1$, and the denominator df $= n-k$. Once the null hypothesis is rejected, it may be desired to estimate all the treatment effects simultaneously.

Simultaneous Confidence Intervals for the Treatment Means

It can be shown that

$$\sqrt{n_1}(\bar{X}_1 - \mu_1)/\sigma, \ldots, \sqrt{n_k}(\bar{X}_k - \mu_k)/\sigma$$

are independent standard normal random variables, and they are independent of

$$\frac{(n-k)S_p^2}{\sigma^2} \sim \chi^2_{n-k}.$$

Define

$$Y = \max_{1 \le i \le k} \left\{ \frac{\sqrt{n_i}|(\bar{X}_i - \mu_i)|}{S_p} \right\}.$$

Then, Y is distributed as X in (13.4.1). Thus, if c denotes the $(1-\alpha)$th quantile of Y, then

$$\bar{X}_1 \pm c\frac{S_p}{\sqrt{n_1}}, \ldots, \bar{X}_k \pm c\frac{S_p}{\sqrt{n_k}} \tag{13.4.2}$$

are exact simultaneous confidence intervals for $\mu_1, \ldots, \mu_k$.

13.4.2 Computing Table Values

The dialog box [StatCalc→Continuous→Student's t→Distribution of max$\{|t_1|, ..., |t_k|\}$] computes the cumulative probabilities and the percentiles of X defined in (13.4.1).

To compute probabilities: Enter the values of the number of groups k, df, and the observed value x of X defined in (13.4.1); click [P(X <= x)].

Example 13.4.1 When $k = 4$, df $= 45$ and $x = 2.3$, $P(X \le 2.3) = 0.900976$ and $P(X > 2.3) = 0.099024$.

To compute percentiles: Enter the values of k, df, and the cumulative probability; click [x].

Example 13.4.2 When $k = 4$, df $= 45$, and the cumulative probability is 0.95, the 95th percentile is 2.5897. That is, $P(X \le 2.5897) = 0.95$.

13.4.3 An Example

Example 13.4.3 Consider the one-way ANOVA model with the following summary statistics:

treatments	1	2	3
sample sizes	11	9	14
sample means	5	3	7
sample variances	4	3	6

The pooled variance S_p^2 is computed as 4.58. Let us compute 95% simultaneous confidence intervals for the mean treatment effects. To get the critical point using *StatCalc*, select the dialog box [StatCalc→Continuous→Student's t→Distribution of $\max\{|t_1|, ..., |t_k|\}$], enter 3 for k, 11 + 9 + 14 - 3 = 31 for df, 0.95 for [P(X <= x)], and click [x]. The required critical point is 2.5178, and the 95% simultaneous confidence intervals for the mean treatment effects based on (13.4.2) are

$$5 \pm 2.5178\sqrt{\frac{4.58}{11}}, \quad 3 \pm 2.5178\sqrt{\frac{4.58}{9}}, \quad 7 \pm 2.5178\sqrt{\frac{4.58}{14}}.$$

13.5 Properties and Results

13.5.1 Properties

1. The t distribution is symmetric about 0. That is,

$$P(-x \le t < 0) = P(0 < t \le x).$$

2. Let X and Y be independent chi-square random variables with degrees of freedoms 1 and n, respectively. Let I be a random variable independent of X and Y such that $P(I = 1) = P(I = -1) = 1/2$. Then

$$I\sqrt{\frac{X}{Y/n}} \sim t_n.$$

3. If X and Y are independent chi-square random variables with df $= n$, then

$$\frac{0.5\sqrt{n}(X - Y)}{\sqrt{XY}} \sim t_n.$$

13.5.2 Relation to Other Distributions

1. Let $F_{1,n}$ denote the F random variable with the numerator df $= 1$, and the denominator df $= n$. Then, for any $x > 0$,

   a. $P(t_n^2 \le x) = P(F_{1,n} \le x)$

   b. $P(F_{1,n} \le x) = 2P(t_n \le \sqrt{x}) - 1$

   c. $P(t_n \le x) = \frac{1}{2}\left[P(F_{1,n} \le x^2) + 1\right]$

2. Let $t_{n,\alpha}$ denote the αth quantile of Student's t distribution with df $= n$. Then

   a. $F_{n,n,\alpha} = 1 + \frac{2(t_{n,\alpha})^2}{n} + \frac{2t_{n,\alpha}}{\sqrt{n}}\sqrt{1 + \frac{(t_{n,\alpha})^2}{n}}$

   b. $t_{n,\alpha} = \frac{\sqrt{n}}{2}\left(\frac{F_{n,n,\alpha}-1}{\sqrt{F_{n,n,\alpha}}}\right)$.

[Cacoullos 1965]

3. Relation to beta distribution: (see Section 16.6.2)

13.5.3 Series Expansions for Cumulative Probability

1. For odd n,

$$P(t_n \le x) = 0.5 + \frac{\arctan(c)}{\pi} + \frac{cd}{\pi}\sum_{k=0}^{(n-3)/2} a_k d^k,$$

and for even n,

$$P(t_n \le x) = 0.5 + \frac{0.5c\sqrt{d}}{\pi}\sum_{k=0}^{(n-2)/2} b_k d^k,$$

where

$$\begin{array}{l} a_0 = 1, \quad b_0 = 1, \\ a_k = \frac{2ka_{k-1}}{2k+1}, \quad b_k = \frac{(2k-1)b_{k-1}}{2k}, \\ c = x/\sqrt{n}, \text{ and } d = \frac{n}{n+x^2}. \end{array}$$

[Owen 1968]

2. Let $x = \arctan(t/\sqrt{n})$. Then, for $n > 1$ and odd,

$$\begin{aligned} P(|t_n| \le t) &= \frac{2}{\pi}\left[x + \sin(x)\left(\cos(x) + \frac{2}{3}\cos^3(x) + \ldots\right.\right. \\ &+ \left.\left.\frac{2 \cdot 4 \cdot \ldots \cdot (n-3)}{1 \cdot 3 \cdot \ldots \cdot (n-2)}\cos^{n-2}(x)\right)\right] \end{aligned}$$

for even n,

$$\begin{aligned} P(|t_n| \le t) &= \sin(x)\left[1 + \frac{1}{2}\cos^2(x) + \frac{1 \cdot 3}{2 \cdot 4}\cos^4(x) + \ldots\right. \\ &+ \left.\frac{1 \cdot 3 \cdot 5 \ldots (n-3)}{2 \cdot 4 \cdot 6 \ldots (n-2)}\cos^{n-2}(x)\right], \end{aligned}$$

and $P(|t_1| \le t) = \frac{2x}{\pi}$. [Abramowitz and Stegun 1965, p. 948]

13.5.4 An Approximation

$$P(t_n \leq t) \simeq P\left(Z \leq \frac{t\left(1 - \frac{1}{4n}\right)}{\sqrt{1 + \frac{t^2}{2n}}}\right),$$

where Z is the standard normal random variable.

13.6 Random Number Generation

Algorithm 13.5.1

Generate Z from $N(0, 1)$
Generate S from gamma$(n/2, 2)$
Set $x = \frac{Z}{\sqrt{S/n}}$.

Then, x is a Student's t random variate with df $= n$.

13.7 A Computational Method for Probabilities

For small integer degrees of freedoms, the series expansions in Section 13.4 can be used to compute the cumulative probabilities. For other degrees of freedoms, use the relation that, for $x > 0$,

$$P(t_n \leq x) = \frac{1}{2}\left[P\left(Y \leq \frac{x^2}{n + x^2}\right) + 1\right],$$

where Y is a beta$(1/2, n/2)$ random variable. If x is negative, then $P(t_n \leq x) = 1 - P(t_n \leq y)$, where $y = -x$.

Chapter 14

Exponential Distribution

14.1 Description

A classical situation in which an exponential distribution arises is as follows: Consider a Poisson process with mean λ where we count the events occurring in a given interval of time or space. Let X denote the waiting time until the first event to occur. Then, for a given $x > 0$,

$$\begin{aligned} P(X > x) &= P(\text{no event in } (0, x)) \\ &= \exp(-x\lambda), \end{aligned}$$

and hence

$$P(X \le x) = 1 - \exp(-x\lambda). \qquad (14.1.1)$$

The distribution in (14.1.1) is called the exponential distribution with mean waiting time $b = 1/\lambda$. The probability density function is given by

$$f(x|b) = \frac{1}{b}\exp(-x/b), \quad x > 0,\ b > 0. \qquad (14.1.2)$$

Suppose that the waiting time is known to exceed a threshold value a, then the pdf is given by

$$f(x|a,b) = \frac{1}{b}\exp(-(x-a)/b), \quad x > a,\ b > 0. \qquad (14.1.3)$$

The distribution with the above pdf is called the two-parameter exponential distribution, and we referred to it as exponential(a, b). The cdf is given by

$$F(x|a,b) = 1 - \exp(-(x-a)/b), \quad x > a,\ b > 0. \qquad (14.1.4)$$

14.2 Moments

The following formulas are valid when $a = 0$.

Mean:	b
Variance:	b^2
Mode:	0
Coefficient of Variation:	1
Coefficient of Skewness:	2
Coefficient of Kurtosis:	9
Moment Generating Function:	$(1 - bt)^{-1}, \quad t < \frac{1}{b}.$
Moments about the Origin:	$E(X^k) = \Gamma(k+1)b^k = k!\, b^k, \; k = 1, 2, \ldots$

14.3 Computing Table Values

The dialog box [StatCalc→Continuous→Exponential] computes the probabilities percentiles, moments and other parameters of an exponential distribution.

To compute probabilities: Enter the values of the shape parameter a, scal parameter b and the observed value x; click on [P(X <= x)].

Example 14.3.1 When $a = 1.1$, $b = 1.6$ and $x = 2$, $P(X \le 2) = 0.430217$ an $P(X > 2) = 0.569783$.

To compute percentiles: Enter the values of a, b, and the cumulative probability click [x].

Example 14.3.2 When $a = 2$, $b = 3$ and the cumulative probability $= 0.05$, th 5th percentile is 2.15388. That is, $P(X \le 2.15388) = 0.05$.

To compute other parameters: Enter the values of the cumulative probability one of the parameters, and a positive value for x; click on the parameter that i missing.

Example 14.3.3 When $b = 3$, $x = 7$ and $P(X \leq x) = 0.9$, the value of the location parameter $a = 0.0922454$.

To compute moments: Enter the values of a and b; click [M].

14.4 Inferences

Let $X_1, \ldots, X_n$ be a sample of observations from an exponential distribution with pdf in (14.1.3).

Maximum Likelihood Estimators

The maximum likelihood estimators of a and b are given by

$$\widehat{a} = X_{(1)} \quad \text{and} \quad \widehat{b} = \frac{1}{n}\sum_{i=1}^{n}(X_i - X_{(1)}) = \bar{X} - X_{(1)}, \tag{14.1.5}$$

where $X_{(1)}$ is the smallest of the X_i's. The MLEs $\widehat{a}$ and $\widehat{b}$ are independent with

$$\frac{2n(\widehat{a} - a)}{b} \sim \chi^2_2 \quad \text{and} \quad \frac{2n\widehat{b}}{b} \sim \chi^2_{2n-2}. \tag{14.1.6}$$

[see Lawless (1982), Section 3.5]

Confidence Intervals

The pivotal quantity $2n\widehat{b}/b$ in (14.1.6) can be used to make inference on b. In particular, a $1 - \alpha$ confidence interval for b is given by

$$\left(\frac{2n\widehat{b}}{\chi^2_{2n-2,1-\alpha/2}}, \frac{2n\widehat{b}}{\chi^2_{2n-2,\alpha/2}}\right).$$

It follows from (14.1.6) that

$$\frac{\widehat{a} - a}{\widehat{b}} \sim \frac{1}{n-1}F_{2,2n-2}.$$

A $1 - \alpha$ confidence interval for a (based on the above distributional result) is given by

$$\left(\widehat{a} - \frac{\widehat{b}}{n-1}F_{2,2n-2,1-\alpha/2},\ \widehat{a} - \frac{\widehat{b}}{n-1}F_{2,2n-2,\alpha/2}\right).$$

14.5 Properties and Results

14.5.1 Properties

1. Memoryless Property: For a given $t > 0$ and $s > 0$,

$$P(X > t + s | X > s) = P(X > t),$$

where X is the exponential random variable with pdf (14.1.3).

2. Let $X_1, \ldots, X_n$ be independent exponential$(0, b)$ random variables. Then

$$\sum_{i=1}^{n} X_i \sim \text{gamma}\,(n, b)\,.$$

3. Let $X_1, \ldots, X_n$ be a sample from an exponential$(0, b)$ distribution. Then, the smallest order statistic $X_{(1)} = \min\{X_1, ..., X_n\}$ has the exponential$(0, b/n)$ distribution.

14.5.2 Relation to Other Distributions

1. Pareto: If X follows a Pareto distribution with pdf $\lambda\sigma^{\lambda}/x^{\lambda+1}$, $x > \sigma$, $\sigma > 0$, $\lambda > 0$, then $Y = \ln(X)$ has the exponential(a, b) distribution with $a = \ln(\sigma)$ and $b = 1/\lambda$.

2. Power Distribution: If X follows a power distribution with pdf $\lambda x^{\lambda-1}/\sigma^{\lambda}$, $0 < x < \lambda$, $\sigma > 0$, then $Y = \ln(1/X)$ has the exponential(a, b) distribution with $a = \ln(1/\sigma)$ and $b = 1/\lambda$.

3. Weibull: See Section 24.6.

4. Extreme Value Distribution: See Section 25.6.

5. Geometric: Let X be a geometric random variable with success probability p. Then

$$P(X \le k | p) = P(Y \le k + 1),$$

where Y is an exponential random variable with mean $b^* = (-\ln(1-p))^{-1}$. [Prochaska 1973]

14.6 Random Number Generation

```
Input:  a = location parameter
        b = scale parameter
Output: x is a random number from the exponential(a, b) distribution

Generate u from uniform(0, 1)
Set x = a - b*ln(u)
```

Chapter 15

Gamma Distribution

15.1 Description

The gamma distribution can be viewed as a generalization of the exponential distribution with mean $1/\lambda$, $\lambda > 0$. An exponential random variable with mean $1/\lambda$ represents the waiting time until the first event to occur, where events are generated by a Poisson process with mean λ, while the gamma random variable X represents the waiting time until the ath event to occur. Therefore,

$$X = \sum_{i}^{a} Y_i,$$

where $Y_1, \ldots, Y_n$ are independent exponential random variables with mean $1/\lambda$. The probability density function of X is given by

$$f(x|a,b) = \frac{1}{\Gamma(a)b^a} e^{-x/b} x^{a-1}, \quad x > 0,\ a > 0,\ b > 0, \tag{15.1.1}$$

where $b = 1/\lambda$. The distribution defined by (15.1.1) is called the gamma distribution with shape parameter a and the scale parameter b. It should be noted that (15.1.1) is a valid probability density function for any $a > 0$ and $b > 0$. The gamma distribution with a positive integer shape parameter a is called the *Erlang Distribution*. If a is a positive integer, then

$$\begin{aligned} F(x|a,b) &= P(\text{waiting time until the } a\text{th event is at most } x \text{ units of time}) \\ &= P(\text{observing at least } a \text{ events in } x \text{ units of time when the} \\ &\qquad \text{mean waiting time per event is } b) \end{aligned}$$

$$
\begin{aligned}
&= P(\text{observing at least } a \text{ events in a Poisson process when} \\
&\qquad \text{the mean number of events is } x/b) \\
&= \sum_{k=a}^{\infty} \frac{e^{-x/b}(x/b)^k}{k!} \\
&= P(Y \geq a),
\end{aligned}
$$

where $Y \sim \text{Poisson}(x/b)$.

The three-parameter gamma distribution has the pdf

$$
f(x|a,b,c) = \frac{1}{\Gamma(a)b^a} e^{-(x-c)/b}(x-c)^{a-1}, \quad a > 0,\ b > 0,\ x > c,
$$

where c is the location parameter. The standard form of gamma distribution (when $b = 1$ and $c = 0$) has the pdf

$$
f(x|a,b) = \frac{1}{\Gamma(a)} e^{-x} x^{a-1}, \quad x > 0,\ a > 0, \tag{15.1.2}
$$

and cumulative distribution function

$$
F(x|a) = \frac{1}{\Gamma(a)} \int_0^x e^{-t} t^{a-1} dt. \tag{15.1.3}
$$

The cdf in (15.1.3) is often referred to as the *incomplete gamma function.*

The gamma probability density plots in Figure 15.1 indicate that the degree of asymmetry of the gamma distribution diminishes as a increases. For large a, $(X - a)/\sqrt{a}$ is approximately distributed as the standard normal random variable.

15.2 Moments

Mean:	ab
Variance:	ab^2
Mode:	$b(a-1), \quad a > 1.$
Coefficient of Variation:	$1/\sqrt{a}$
Coefficient of Skewness:	$2/\sqrt{a}$

Coefficient of Kurtosis:	$3 + 6/a$
Moment Generating Function:	$(1 - bt)^{-a}, \quad t < \frac{1}{b}.$
Moments about the Origin:	$\frac{\Gamma(a+k)b^k}{\Gamma(a)} = b^k \prod_{i=1}^{k} (a + i - 1), \; k = 1, 2, \ldots$

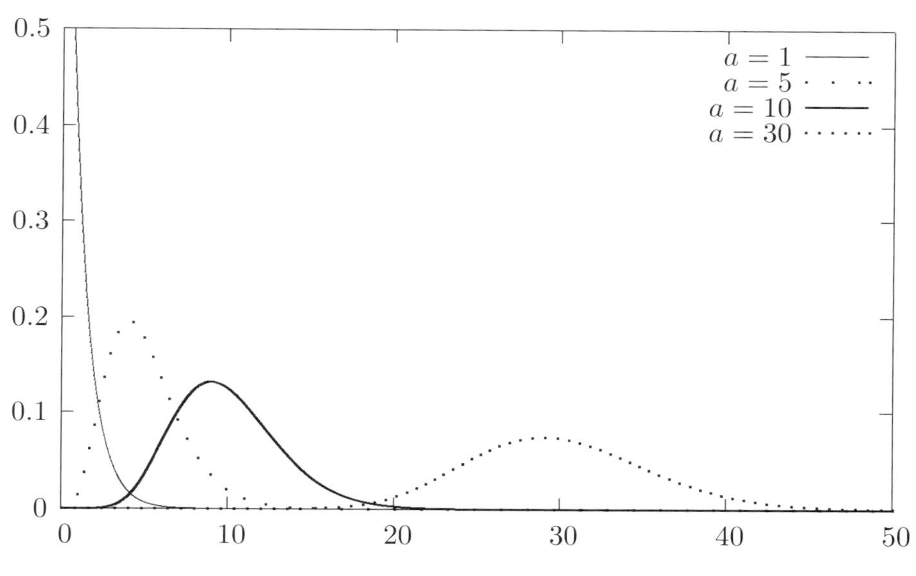

Figure 15.1 Gamma pdfs

15.3 Computing Table Values

The dialog box [StatCalc→Continuous→ Gamma] computes probabilities, percentiles, moments and also the parameters when other values are given.

To compute probabilities: Enter the values of the shape parameter a, scale parameter b and the observed value x; click on [P(X <= x)].

Example 15.3.1 When $a = 2$, $b = 3$ and $x = 5.3$, $P(X \leq 5.3) = 0.527172$ and $P(X > 5.3) = 0.472828$.

To compute percentiles: Enter the values of a, b, and the cumulative probability; click [x].

Example 15.3.2 When $a = 2$, $b = 3$ and the cumulative probability $= 0.05$, the

5th percentile is 1.06608. That is, $P(X \leq 1.06608) = 0.05$.

To compute other Parameters: Enter the values of the probability, one of the parameters, and a positive value for x; click on the parameter that is missing.

Example 15.3.3 When $b = 3$, $x = 5.3$ and $P(X \leq x) = 0.9$, the value of the shape parameter $a = 0.704973$.

To compute moments: Enter the values of a and b; click [M].

15.4 Applications with Some Examples

The gamma distribution arises in situations where one is concerned about the waiting time for a finite number of independent events to occur, assuming that events occur at a constant rate and chances that more than one event occurs in a small interval of time are negligible. This distribution has applications in reliability and queuing theory. Examples include the distribution of failure times of components, the distribution of times between calibration of instruments which need re-calibration after a certain number of uses and the distribution of waiting times of k customers who will arrive at a store. The gamma distribution can also be used to model the amounts of daily rainfall in a region. For example the data on daily rainfall in Sydney, Australia, (October 17 – November 7; years 1859 – 1952) were modeled by a gamma distribution. A gamma distribution was postulated because precipitation occurs only when water particles can form around dust of sufficient mass, and the waiting time for such accumulation of dust is similar to the waiting time aspect implicit in the gamma distribution (Das 1955). Stephenson et al. (1999) showed that the gamma and Weibull distributions provide good fits to the wet-day rainfall distribution in India.

Example 15.4.1 The distribution of fifty-year summer rainfall (in inches) in a certain part of India is approximately gamma with $a = 3.0$ and $b = 2.0$.

a. Find the percentage of summer rainfalls exceed six inches.

b. Find an interval that will contain 95% of the summer rainfall totals.

Solution: Let X denote the total summer rainfall in a year.

a. Select the dialog box [StatCalc→Continuous→ Gamma] from *StatCalc*, enter 3 for a, 2 for b, and 6 for observed x; click [P(X <= x)] to get $P(X > 6) = 0.42319$. That is, about 42% of the summer rainfall totals exceed 6 inches.

b. To find a right endpoint, enter 3 for a, 2 for b, and 0.975 for cumulative probability; click [x] to get 16.71. To find a lower endpoint, enter 0.025 for the cumulative probability and click [x] to get 0.73. Thus, 95% of the summer rainfall totals are between 0.73 and 16.71 inches.

Example 15.4.2 Customers enter a fast food restaurant, according to a Poisson process, on average 4 for every 3-minute period during the peak hours 11:00 am – 1:00 pm Let X denote the waiting time in minutes until arrival of the 60th customer.

a. Find $E(X)$.

b. Find $P(X > 50)$.

Solution: The mean number of customers per minute is 4/3. Therefore, mean waiting time in minutes is $b = 3/4$.

a. $E(X) = ab = 60 \text{ x } 3/4 = 45$ min.

b. To find the probability using [StatCalc→Continuous→ Gamma], enter 60 for a, $3/4 = 0.75$ for b, and 50 for x; click [P(X <= x)] to get $P(X > 50) = 0.19123$.

15.5 Inferences

Let $X_1, \ldots, X_n$ be a sample from a gamma distribution with the shape parameter a, scale parameter b, and the location parameter c. Let $\bar{X}$ denote the sample mean.

15.5.1 Maximum Likelihood Estimators

The MLEs of a, b and c are the solutions of the equations

$$\begin{aligned} \sum_{i=1}^{n} \ln(X_i - c) - n \ln b - n\psi(a) &= 0 \\ \sum_{i=1}^{n} (X_i - c) - nab &= 0 \\ \sum_{i=1}^{n} (X_i - c)^{-1} + n[b(a-1)]^{-1} &= 0 \end{aligned} \qquad (15.5.1)$$

where ψ is the digamma function (see Section 1.8). These equations may yield reliable solutions if a is expected to be at least 2.5.

If the location parameter c is known, the MLEs of a and b are the solutions of the equations

$$\frac{1}{n}\sum_{i=1}^{n}\ln(X_j - c) - \ln(\bar{X} - c) - \psi(a) + \ln a = 0 \quad \text{and} \quad ab = \bar{X}.$$

If a is also known, then $\bar{X}/a$ is the UMVUE of b.

15.5.2 Moment Estimators

Moment estimators are given by

$$\widehat{a} = \frac{4m_2^3}{m_3^2}, \quad \widehat{b} = \frac{m_3}{2m_2} \quad \text{and} \quad \widehat{c} = \bar{X} - 2\frac{m_2^2}{m_3},$$

where

$$m_k = \frac{1}{n}\sum_{i=1}^{n}(X_i - \bar{X})^k, \quad k = 1, 2, \ldots$$

is the kth sample central moment.

15.5.3 Interval Estimation

Let a be known and $S = n\bar{X}$. Let S_0 be an observed value of S. The endpoints of a $1-\alpha$ confidence interval (b_L, b_U) satisfy

$$P(S \le S_0 | b_U) = \alpha/2, \tag{15.5.2}$$

and

$$P(S \ge S_0 | b_L) = \alpha/2. \tag{15.5.3}$$

Since $S \sim$ gamma(na, b), it follows from (15.5.2) and (15.5.3) that

$$(b_L, b_U) = \left(\frac{S_0}{\text{gamma}^{-1}(1-\alpha/2; na, 1)}, \frac{S_0}{\text{gamma}^{-1}(\alpha/2; na, 1)}\right),$$

where gamma$^{-1}(p; d, 1)$ denotes the pth quantile of a gamma distribution with the shape parameter d and scale parameter 1, is a $1-\alpha$ confidence interval for b [Guenther 1969 and 1971].

The dialog box [StatCalc→Continuous→Gamma→CI for b] uses the above formula to compute confidence intervals for b.

Example 15.5.1 Suppose that a sample of 10 observations from a gamma population with shape parameter $a = 1.5$ and unknown scale parameter b produced a mean value of 2. To find a 95% confidence interval for b, enter these values in appropriate edit boxes, and click [2-sided] to get (0.85144, 2.38226). To get one-sided limits, click [1-sided] to get 0.913806 and 2.16302. This means that the true value of b is at least 0.913806 with confidence 0.95; the true value of b is at most 2.16302 with confidence 0.95.

Suppose we want to test

$$H_0 : b \le 0.7 \quad \text{vs.} \quad H_a : b > 0.7.$$

To get the p-value, enter 0.7 for [H0: b = b0] and click [p-values for] to get 0.00201325. Thus, we conclude that b is significantly greater than 0.7.

15.6 Properties and Results

1. An Identity: Let $F(x|a,b)$ and $f(x|a,b)$ denote, respectively, the cdf and pdf of a gamma random variable X with parameters a and b. Then,

$$F(x|a,\, 1) = F(x|a+1, 1) + f(x|a+1, 1).$$

2. Additive Property: Let $X_1, \ldots, X_k$ be independent gamma random variables with the same scale parameter but possibly different shape parameters $a_1, \ldots, a_k$, respectively. Then

$$\sum_{i=1}^{k} X_i \sim \text{gamma}\left(\sum_{i=1}^{k} a_i, b\right).$$

3. Exponential: Let $X_1, \ldots, X_n$ be independent exponential random variables with mean b. Then

$$\sum_{i=1}^{n} X_i \sim \text{gamma}(n,\, b).$$

4. Chi-square: When $a = n/2$ and $b = 2$, the gamma distribution specializes to the chi-square distribution with df $= n$.

5. Beta: See Section 16.6.

6. Student's t: If X and Y are independent gamma(n, 1) random variables, then
$$\sqrt{n/2}\left(\frac{X-Y}{\sqrt{XY}}\right) \sim t_{2n}.$$

15.7 Random Number Generation

```
Input:  a = shape parameter gamma(a) distribution
Output: x = gamma(a) random variate
        y = b*x is a random number from gamma(a, b).
```

Algorithm 15.7.1

```
For a = 1:

Generate u from uniform(0, 1) return x = -ln(u)
```

The following algorithm for $a > 1$ is due to Schmeiser and Lal (1980). When $0 < a < 1$, X = gamma(a) variate can be generated using relation that $X = U^{\frac{1}{a}}Z$, where Z is a gamma($a+1$) random variate.

[1]*Algorithm 15.7.2*

```
      Set f(x) = exp(x3*ln(x/x3) + x3 - x)
      x3 = a-1
      d = sqrt(x3)
      k =1
      x1 = x2 = f2 = 0
      If d >= x3, go to 2
      x2 = x3 - d
      k = 1- x3/x2
      x1 = x2 + 1/k
      f2 = f(x2)

2     Set x4 = x3 + d
          r = 1 - x3/x4
          x5 = x4 + 1/r
          f4 = f(x4)
```

[1]Reproduced with permission from the American Statistical Association.

```
        p1 = x4 - x2
        p2 = p1 - f2/k
        p3 = p2 + f4/r

3     Generate u, v from uniform(0, 1)
      Set u = u*p3
      If u > p1 go to 4
      Set x = x2 + u
      If x > x3 and v <= f4 + (x4 - x)*(1 - f4)/(x4 - x3), return x
      If x < x3 and v <= f2 + (x - x2)*(1 - f2)/(x3 - x2), return x
      go to 6

4     If u > p2, go to 5
      Set u = (u - p1)/(p2 - p1)
      x = x2 - ln(u)/k
      If x < 0, go to 3
      Set v = v*f2*u
      If v <= f2*(x - x1)/(x2 - x1) return x
      go to 6
5     Set u = (u - p2)/(p3 - p2)
      x = x4 - ln(u)/r
      v = v*f4*u
      If v <= f4*(x5 - x)/(x5 - x4) return x

6     If ln(v) <= x3*ln(x/x3) + x3 - x, return x
      else go to 3
```

x is a random number from the gamma(a, 1) distribution.

15.8 A Computational Method for Probabilities

To compute $P(X \leq x)$ *when* $a > 0$ *and* $b = 1$:

The Pearson's series for the cdf is given by

$$P(X \leq x) = \exp(-x)x^a \sum_{i=0}^{\infty} \frac{1}{\Gamma(a+1+i)} x^i. \tag{15.8.1}$$

The cdf can also be computed using the continued fraction:

$$P(X > x) = \frac{\exp(-x)x^a}{\Gamma(a)} \left(\frac{1}{x+1-a-} \, \frac{1 \cdot (1-a)}{x+3-a-} \, \frac{2 \cdot (2-a)}{x+5-a-} \cdots \right) \quad (15.8.2)$$

To compute $\Gamma(a+1)$ use the relation $\Gamma(a+1) = a\Gamma(a)$ [Press et al. 1992].

The series (15.8.1) converges faster for $x < a+1$ while the continued fraction (15.8.2) converges faster for $x \geq a+1$. A method of evaluating continued fraction is given in Kennedy and Gentle (1980, p. 76).

The following Fortran function routine is based on the series expansion of the cdf in (15.8.1).

```
Input:
      x = the value at which the cdf is to be evaluated, x > 0
      a = shape parameter > 0

Output:
      P(X <= x) = gamcdf(x, a)

cccccccccccccccccccccccccccccccccccccccccccccccccccccccccccccccc
     doubleprecision function gamcdf(x, a)
     implicit doubleprecision(a-h, o-z)
     data one, maxitr, error/1.0d0, 1000, 1.0d-12/

c Logarithmic gamma function alng(x) in Section 1.8.1 is required

     com  = dexp(a*dlog(x)-alng(a)-x);
     a0 = a;
     term = one/a;  sum = one/a;

     do i = 1, maxitr
       a0 = a0 + one
       term = term*x/a0;
       sum = sum + term;
       if (dabs(term) < sum*error) goto 1
     end do

1    gamcdf = sum*com
     end
```

Chapter 16

Beta Distribution

16.1 Description

The probability density function of a beta random variable with shape parameters a and b is given by

$$f(x|a,b) = \frac{1}{B(a,b)}\, x^{a-1}(1-x)^{b-1}, \qquad 0 < x < 1,\ a > 0,\ b > 0,$$

where the beta function $B(a, b) = \Gamma(a)\Gamma(b)/\Gamma(a+b)$. We denote the above beta distribution by beta(a, b). A situation where the beta distribution arises is given below.

Consider a Poisson process with arrival rate of λ events per unit time. Let W_k denote the waiting time until the kth arrival of an event and W_s denote the waiting time until the sth arrival, $s > k$. Then, W_k and $W_s - W_k$ are independent gamma random variables with

$$W_k \sim \text{gamma}(k, 1/\lambda) \text{ and } W_s - W_k \sim \text{gamma}(s-k, 1/\lambda).$$

The proportion of the time taken by the first k arrivals in the time needed for the first s arrivals is

$$\frac{W_k}{W_s} = \frac{W_k}{W_k + (W_s - W_k)} \sim \text{beta}(k, s-k).$$

The beta density plots are given for various values of a and b in Figure 16.1. We observe from the plots that the beta density is U shaped when $a < 1$ and

$b < 1$, symmetric about 0.5 when $a = b > 1$, J shaped when $(a-1)(b-1) < 0$, and unimodal for other values of a and b. For equally large values of a and b, the cumulative probabilities of a beta distributions can be approximated by a normal distribution.

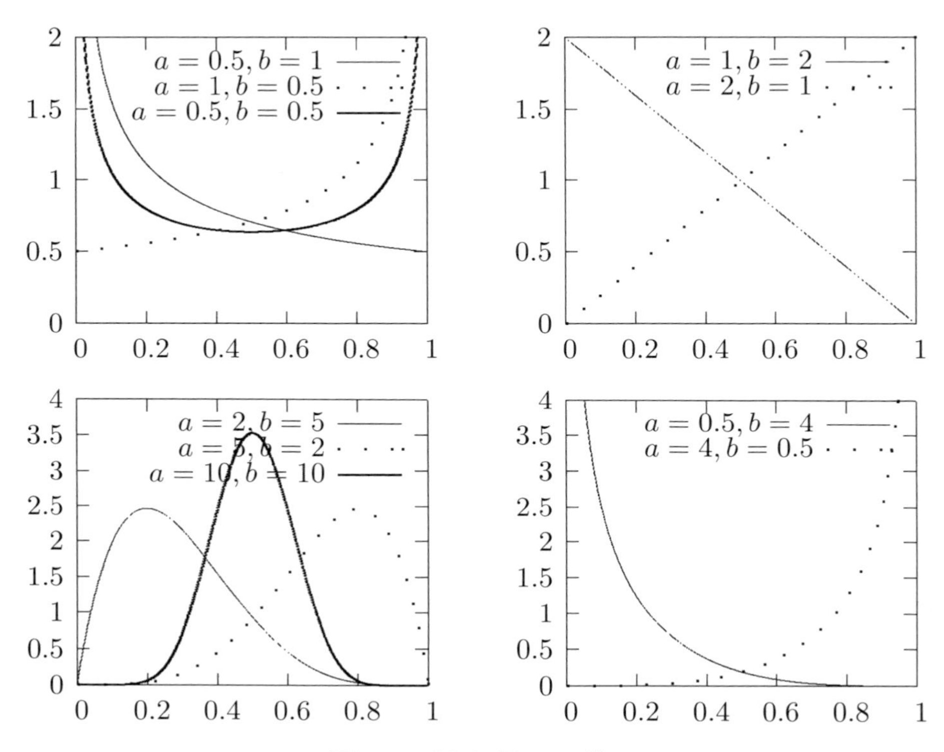

Figure 16.1 Beta pdfs

16.2 Moments

Mean:	$\frac{a}{a+b}$
Variance:	$\frac{ab}{(a+b)^2(a+b+1)}$
Mode:	$\frac{a-1}{a+b-2}, \quad a > 1, \ b > 1.$
Mean Deviation:	$\frac{\Gamma(a+b)}{\Gamma(a)\Gamma(b)} \frac{2a^a b^b}{(a+b)^{(a+b+1)}}$

Coefficient of Skewness:	$\frac{2(b-a)(a+b+1)^{1/2}}{(a+b+2)(ab)^{1/2}}$
Coefficient of Variation:	$\frac{\sqrt{b}}{\sqrt{a(a+b+1)}}$
Coefficient of Kurtosis:	$\frac{3(a+b+1)[2(a+b)^2+ab(a+b-6)]}{ab(a+b+2)(a+b+3)}$
Characteristic Function:	$\frac{\Gamma(a+b)}{\Gamma(a)} \sum_{k=0}^{\infty} \frac{\Gamma(a+k)(it)^2}{\Gamma(a+b+k)\Gamma(k+1)}$
Moments about the Origin:	$E(X^k) = \prod_{i=0}^{k-1} \frac{a+i}{a+b+i}, \quad k = 1, 2, \ldots$

16.3 Computing Table Values

The dialog box [StatCalc→Continuous→Beta] computes the cdf, percentiles and moments of a beta distribution.

To compute probabilities: Enter the values of the parameters a and b, and the value of x; click [P(X <= x)].

Example 16.3.1 When $a = 2$, $b = 3$, and $x = 0.4$, $P(X \leq 0.4) = 0.5248$ and $P(X > 0.4) = 0.4752$.

To compute percentiles: Enter the values of a, b and the cumulative probability; click [x].

Example 16.3.2 When $a = 2$, $b = 3$, and the cumulative probability $= 0.40$, the 40th percentile is 0.329167. That is, $P(X \leq 0.329167) = 0.40$.

To compute other parameters: Enter the values of one of the parameters, cumulative probability, and the value of x; click on the missing parameter.

Example 16.3.3 When $b = 3$, $x = 0.8$, and the cumulative probability $= 0.40$, the value of a is 12.959.

To compute moments: Enter the values of a and b and click [M].

16.4 Inferences

Let $X_1, \ldots, X_n$ be a sample from a beta distribution with shape parameters a and b. Let

$$\bar{X} = \frac{1}{n}\sum_{i=1}^{n} X_i \text{ and } S^2 = \frac{1}{n-1}\sum_{i=1}^{n}(X_i - \bar{X})^2.$$

Moment Estimators

$$\hat{a} = \bar{X}\left[\frac{\bar{X}(1-\bar{X})}{S^2} - 1\right]$$

and

$$\hat{b} = \frac{(1-\bar{X})\hat{a}}{\bar{X}}.$$

Maximum Likelihood Estimators

MLEs are the solution of the equations

$$\begin{aligned}
\psi(\hat{a}) - \psi(\hat{a}+\hat{b}) &= \frac{1}{n}\sum_{i=1}^{n}\ln(X_i) \\
\psi(\hat{b}) - \psi(\hat{a}+\hat{b}) &= \frac{1}{n}\sum_{i=1}^{n}\ln(1-X_i),
\end{aligned}$$

where $\psi(x)$ is the digamma function given in Section 1.8. Moment estimators can be used as initial values to solve the above equations numerically.

16.5 Applications with an Example

As mentioned in earlier chapters, the beta distribution is related to many other distributions such as Student's t, F, noncentral F, binomial and negative binomial distributions. Therefore, cumulative probabilities and percentiles of these distributions can be obtained from those of beta distributions. For example, as mentioned in Sections 3.5 and 7.6, percentiles of beta distributions can be used to construct exact confidence limits for binomial and negative binomial success probabilities. In Bayesian analysis, the beta distribution is considered as a conjugate prior distribution for the binomial success probability p. Beta distributions are often used to model data consisting of proportions. Applications of beta distributions in risk analysis are mentioned in Johnson (1997).

Chia and Hutchinson (1991) used a beta distribution to fit the frequency distribution of daily cloud durations, where cloud duration is defined as the fraction of daylight hours not receiving bright sunshine. They used data collected from 11 Australian locations to construct 132 (11 stations by 12 months) empirical frequency distributions of daily cloud duration. Sulaiman et al. (1999) fitted Malaysian sunshine data covering a 10-year period to a beta distribution. Nicas (1994) pointed out that beta distributions offer greater flexibility than lognormal distributions in modeling respirator penetration values over the physically plausible interval [0,1]. An approach for dynamically computing the retirement probability and the retirement rate when the age manpower follows a beta distribution is given in Shivanagaraju et al. (1998). The coefficient of kurtosis of the beta distribution has been used as a good indicator of the condition of a gear (Oguamanam et al. 1995). SchwarzenbergCzerny (1997) showed that the phase dispersion minimization statistic (a popular method for searching for nonsinusoidal pulsations) follows a beta distribution. In the following we give an illustrative example.

Example 16.5.1 National Climatic Center (North Carolina, USA) reported the following data in Table 16.1 on percentage of day during which sunshine occurred in Atlanta, Georgia, November 1–30, 1974. Daniel (1990) considered these data to demonstrate the application of a run test for testing randomness. We will fit a beta distribution for the data.

Table 16.1 Percentage of sunshine period in a day in November 1974

85	85	99	70	17	74	100	28	100	100	31	86	100	0	100
100	45	7	12	54	87	100	100	88	50	100	100	100	48	0

To fit a beta distribution, we first compute the mean and variance of the data:

$$\bar{x} = 0.6887 \quad \text{and} \quad s^2 = 0.1276.$$

Using the computed mean and variance, we compute the moment estimators (see Section 16.4) as

$$\hat{a} = 0.4687 \quad \text{and} \quad \hat{b} = 0.2116.$$

The observed quantiles q_j (that is, the ordered proportions) for the data are given in the second column of Table 16.2. The estimated shape parameters can be used to compute the beta quantiles so that they can be compared with the corresponding observed quantiles. For example, when the observed quantile is 0.31 (at $j = 7$), the corresponding beta quantile Q_j can be computed as

$$Q_j = \text{beta}^{-1}(0.21667; \hat{a}, \hat{b}) = 0.30308,$$

where $\text{beta}^{-1}(p; \hat{a}, \hat{b})$ denotes the $100p$th percentile of the beta distribution with shape parameters $\hat{a}$ and $\hat{b}$. Comparison between the sample quantiles and the corresponding beta quantiles (see the Q–Q plot in Figure 16.1) indicates that the data set is well fitted by the beta($\hat{a}$, $\hat{b}$)distribution. Using this fitted beta distribution, we can estimate the probability that the sunshine period exceeds a given proportion in a November day in Atlanta. For example, the estimated probability that at least 70% of a November day will have sunshine is given by $P(X \geq 0.7) = 0.61546$, where X is the beta(0.4687, 0.2116) random variable.

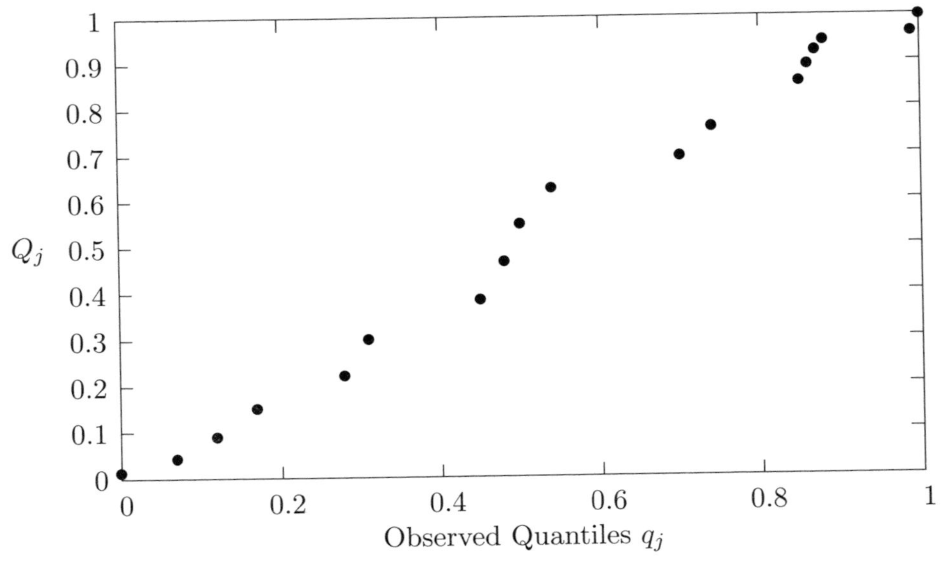

Figure 16.1 Q-Q Plots of the Sunshine Data

Table 16.2 Observed and beta quantiles for sunshine data

j	Observed Quantiles q_j	Cumulative Probability Levels $(j - 0.5)/30$	Beta Quantiles Q_j
1	0		
2	0	0.05	0.01586
3	0.7	0.083333	0.04639
4	0.12	0.116667	0.09263
5	0.17	0.15	0.15278
6	0.28	0.183333	0.22404
7	0.31	0.216667	0.30308
8	0.45	0.25	0.38625
9	0.48	0.283333	0.47009
10	0.5	0.316667	0.55153
11	0.54	0.35	0.62802
12	0.7	0.383333	0.69772
13	0.74	0.416667	0.75946
14	0.85		
15	0.85	0.483333	0.85746
16	0.86	0.516667	0.89415
17	0.87	0.55	0.92344
18	0.88	0.583333	0.94623
19	0.99	0.616667	0.96346
20	1	...	...
...	...	...	...
30	1	0.983333	1

16.6 Properties and Results

16.6.1 An Identity and Recurrence Relations

1. Let $F(x|a, b)$ denote the cumulative distribution of a beta(a, b) random variable; that is $F(x|a, b) = P(X \leq x|a, b)$.

   a. $F(x|a, b) = 1 - F(1 - x|b, a)$.

   b. $F(x|a, b) = xF(x|a - 1, b) + (1 - x)F(x|a, b - 1), \quad a > 1, b > 1$.

   c. $F(x|a, b) = [F(x|a + 1, b) - (1 - x)F(x|a + 1, b - 1)]/x, \quad b > 1$.

   d. $F(x|a, b) = [aF(x|a + 1, b) + bF(x|a, b + 1)]/(a + b)$.

   e. $F(x|a, b) = \frac{\Gamma(a+b)}{\Gamma(a+1)\Gamma(b)} x^a (1 - x)^{b-1} + F(x|a + 1, b - 1), \quad b > 1$.

f. $F(x|a,b) = \frac{\Gamma(a+b)}{\Gamma(a+1)\Gamma(b)} x^a (1-x)^b + F(x|a+1,b)$.

g. $F(x|a,a) = \frac{1}{2} F(1 - 4(x-0.5)^2 | a, 0.5), \quad x \le 0.5$.

[Abramowitz and Stegun 1965, p. 944]

16.6.2 Relation to Other Distributions

1. Chi-square Distribution: Let X and Y be independent chi-square random variables with degrees of freedom m and n, respectively. Then

$$\frac{X}{X+Y} \sim \text{beta}(m/2,\ n/2) \text{ distribution.}$$

2. Student's t Distribution: Let t be a Student's t random variable with df $= n$. Then

$$P(|t| \le x) = P(Y \le x^2/(n+x^2)) \quad \text{for } x > 0,$$

where Y is a beta(1/2, $n/2$) random variable.

3. Uniform Distribution: The beta(a, b) distribution specializes to the uniform(0,1) distribution when $a = 1$ and $b = 1$.

4. Let $X_1, \ldots, X_n$ be independent uniform(0,1) random variables, and let $X_{(k)}$ denote the kth order statistic. Then, $X_{(k)}$ follows a beta$(k, n-k+1)$ distribution.

5. F Distribution: Let X be a beta($m/2$, $n/2$) random variable . Then

$$\frac{nX}{m(1-X)} \sim F_{m,n} \text{ distribution.}$$

6. Binomial: Let X be a binomial(n, p) random variable. Then, for a given k,

$$P(X \ge k|n,p) = P(Y \le p),$$

where Y is a beta(k, $n-k+1$) random variable. Furthermore,

$$P(X \le k|n,p) = P(W \ge p),$$

where W is a beta(k+ 1, $n-k$) random variable.

7. Negative Binomial: Let X be a negative binomial(r, p) random variable.

$$P(X \leq k | r, p) = P(W \leq p),$$

where W is a beta random variable with parameters r and $k + 1$.

8. Gamma: Let X and Y be independent gamma random variables with the same scale parameter b, but possibly different shape parameters a_1 and a_2. Then

$$\frac{X}{X+Y} \sim \text{beta}(a_1, a_2).$$

16.7 Random Number Generation

The following algorithm generates beta(a, b) variates. It uses the approach by Jöhnk (1964) when $\min\{a, b\} < 1$ and Algorithm 2P of Schmeiser and Shalaby (1980) otherwise.

[1] *Algorithm 16.7.1*

```
Input:
    a, b = the shape parameters

Output:
    x is a random variate from beta(a, b) distribution

      if a > 1 and b > 1, goto 1
2     Generate u1 and u2 from uniform(0, 1)
      Set s1 = u1**(1./a)
          s2 = u2**(1./b)
          s = s1 + s2
          x = s1/s
      if(s <= 1.0) return x
      goto 2

1     Set aa = a - 1.0
          bb = b - 1.0
          r = aa + bb
          s = r*ln(r)
```

[1] Reproduced with permission from the American Statistical Association.

```
        x1 = 0.0; x2 = 0.0
        x3 = aa/r
        x4 = 1.0; x5 = 1.0
        f2 = 0.0; f4 = 0.0
      if(r <= 1.0) goto 4
      d = sqrt(aa*bb/(r-1.0))/r
      if(d >= x3) goto 3
      x2 = x3 - d
      x1 = x2 - (x2*(1.0-x2))/(aa-r*x2)
      f2 = exp(aa*ln(x2/aa)+bb*ln((1.0-x2)/bb)+s)

3     if(x3+d >= 1.0) goto 4
      x4 = x3 + d
      x5 = x4 - (x4*(1.0-x4)/(aa-r*x4))
      f4 = exp(aa*ln(x4/aa) + bb*ln((1.0-x4)/bb)+s)

4     p1 = x3 - x2
      p2 = (x4 - x3) + p1
      p3 = f2*x2/2.0 + p2
      p4 = f4*(1.0-x4)/2.0+ p3

5     Generate u from uniform(0,1)
      Set u = u*p4
      Generate w from uniform(0,1)
      if(u > p1) goto 7
      x = x2 + w*(x3-x2)
      v = u/p1
      if(v <= f2 + w*(1.0-f2)) return x
      goto 10

7     if(u > p2) goto 8
      x = x3 + w*(x4 - x3)
      v = (u - p1)/(p2 - p1)
      if(v <= 1.0 - (1.0-f4)/w) return x
      goto 10

8     Generate w2 from uniform(0,1)
      if(w2 > w) w = w2
      if(u > p3) goto 9
      x = w*x2
      v = (u-p2)/(p3-p2)*w*f2
```

```
      if(x <= x1) goto 10
      if(v <= f2*(x-x1)/(x2-x1)) return x
      goto 10

9     x = 1.0 - w*(1.0-x4)
      v = ((u-p3)/(p4-p3))*(1.0-x)*f4/(1.0-x4)
      if(x >= x5) goto 10
      if(v <= f4*(x5-x)/(x5-x4)) return x

10    ca = ln(v)
      if(ca >= -2.0*r*(x-x3)**2) goto 5
      if(ca <= aa*ln(x/aa)+bb*ln((1.0-x)/bb) + s) return x
      goto 5
```

For other equally efficient algorithms, see Cheng (1978).

16.8 Evaluating the Distribution Function

The recurrence relation

$$F(x|a,b) = \frac{\Gamma(a+b)}{\Gamma(a+1)\Gamma(b)} x^a (1-x)^b + F(x|a+1,b)$$

can be used to evaluate the cdf at a given x, a and b. The above relation produces the series

$$\begin{aligned} F(x|a,b) &= \frac{x^a(1-x)^b}{\text{Beta}(a+1,b)} \left(\frac{1}{a+b} + \frac{x}{a+1} + \frac{(a+b+1)x^2}{(a+1)(a+2)} \right. \\ &+ \left. \frac{(a+b+1)(a+b+2)}{(a+1)(a+2)(a+3)} x^3 + \ldots \right). \end{aligned} \quad (16.1.1)$$

If $x > 0.5$, then, to speed up the convergence, compute first $F(1-x|b,a)$, and then use the relation that $F(x|a,b) = 1 - F(1-x|b,a)$ to evaluate $F(x|a,b)$.

The following Fortran subroutine evaluates the cdf of a beta(a,b) distribution, and is based on the above method.

```
Input:
     x = the value at which the cdf is to be evaluated, x > 0
     a = shape parameter > 0
```

```
      b = shape parameter > 0
Output:
      P(X <= x) = betadf(x, a, b)

cccccccccccccccccccccccccccccccccccccccccccccccccccccccccccccc
    doubleprecision function betadf(x, a, b)
    implicit double precision (a-h,o-z)
    logical check
    data one, error, zero, maxitr/1.0d0, 1.0d-13, 0.0d0, 1000/

    if(x .gt. 0.5d0) then
    xx = one-x; aa = b; bb = a
    check = .true.
    else
    xx = x; aa = a; bb = b
    check = .false.
    end if

    bet = alng(aa+bb+one)-alng(aa+one)-alng(bb)
    sum = zero
    term = xx/(aa+one)
    i = 1
1   sum = sum + term
    if(term .le. error .or. i .ge. maxitr) goto 2
    term = term*(aa+bb+i)*xx/(aa+i+1.0d0)
    i = i + 1
    goto 1

2   betadf = (sum + one/(aa+bb))*dexp(bet+aa*dlog(xx)+bb*dlog(one-
    if(check) betadf = one-betadf
    end
```

Majumder and Bhattacharjee (1973a, Algorithm AS 63) proposed a slightly faster approach than the above method. Their algorithm uses a combination of the recurrence relations 1(e) and 1(f) in Section 16.6.1, depending on the parameter configurations and the value of x at which the cdf is evaluated. For computing percentiles of a beta distribution, see Majumder and Bhattacharjee (1973b, Algorithm AS 64).

Chapter 17

Noncentral Chi-square Distribution

17.1 Description

The probability density function of a noncentral chi-square random variable with the degrees of freedom n and the noncentrality parameter δ is given by

$$f(x|n,\delta) = \sum_{k=0}^{\infty} \frac{\exp\left(-\frac{\delta}{2}\right)\left(\frac{\delta}{2}\right)^k}{k!} \frac{\exp\left(-\frac{x}{2}\right) x^{\frac{n+2k}{2}-1}}{2^{\frac{n+2k}{2}}\Gamma(\frac{n+2k}{2})}, \tag{17.1.1}$$

where $x > 0$, $n > 0$, and $\delta > 0$. This random variable is usually denoted by $\chi_n^2(\delta)$. It is clear from the density function (17.1.1) that conditionally given K, $\chi_n^2(\delta)$ is distributed as χ_{n+2K}^2, where K is a Poisson random variable with mean $\delta/2$. Thus, the cumulative distribution of $\chi_n^2(\delta)$ can be written as

$$P(\chi_n^2(\delta) \le x|n,\delta) = \sum_{k=0}^{\infty} \frac{\exp\left(-\frac{\delta}{2}\right)\left(\frac{\delta}{2}\right)^k}{k!} P(\chi_{n+2k}^2 \le x). \tag{17.1.2}$$

The plots of the noncentral chi-square pdfs in Figure 17.1 show that, for fixed n, $\chi_n^2(\delta)$ is stochastically increasing with respect to δ, and for large values of n, the pdf is approximately symmetric about its mean $n + \delta$.

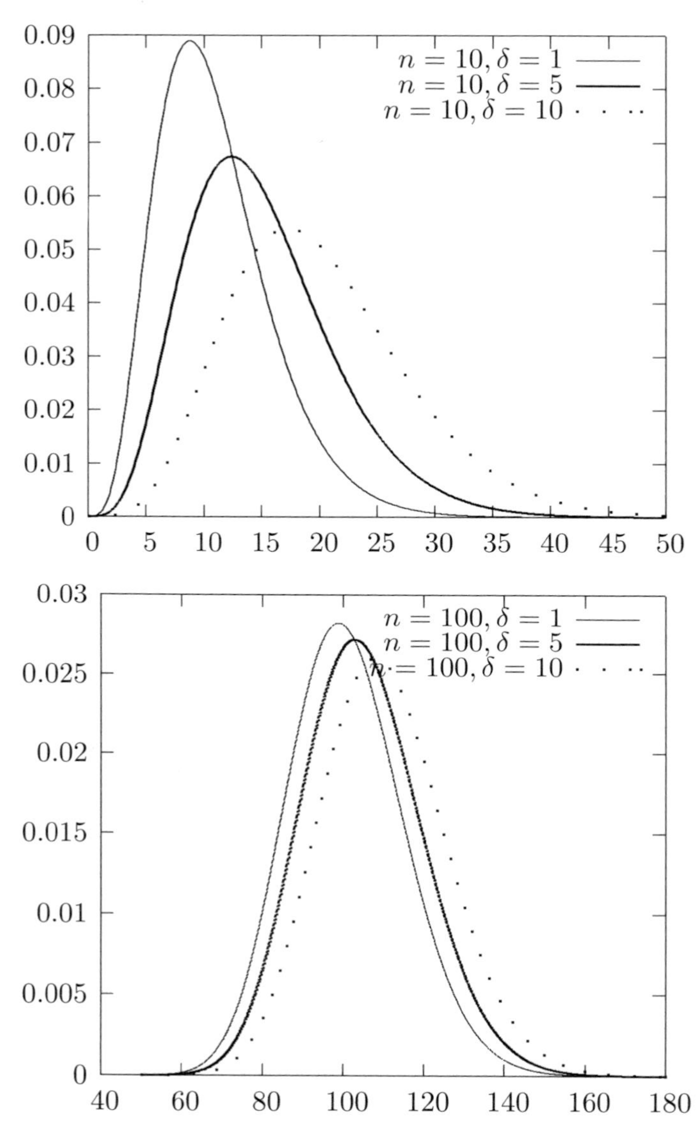

Figure 17.1 Noncentral Chi-square pdfs

17.2 Moments

Mean:	$n + \delta$
Variance:	$2n + 4\delta$
Coefficient of Variation:	$\frac{\sqrt{(2n+4\delta)}}{(n+\delta)}$
Coefficient of Skewness:	$\frac{(n+3\delta)\sqrt{8}}{(n+2\delta)^{3/2}}$
Coefficient of Kurtosis:	$3 + \frac{12(n+4\delta)}{(n+2\delta)^2}$
Moment Generating Function:	$(1-2t)^{-n/2}\exp[t\delta/(1-2t)]$
Moments about the Origin:	$E(X^k) = 2^k\Gamma(n/2+k)\sum_{j=0}^{\infty}\binom{k}{j}\frac{(\delta/2)^j}{\Gamma(n/2+j)}$, $k = 1, 2, \ldots$ [Johnson and Kotz 1970, p. 135]

17.3 Computing Table Values

The dialog box [StatCalc→Continuous→NC Chi-sqr] computes the cdf, percentiles, moments and other parameters of a noncentral chi-square distribution.

To compute probabilities: Enter the values of the df, noncentrality parameter, and the value of x; click [P(X <= x)].

Example 17.3.1 When df = 13.0, noncentrality parameter = 2.2 and the observed value $x = 12.3$,

$$P(X \le 12.3) = 0.346216 \quad \text{and} \quad P(X > 12.3) = 0.653784.$$

To compute percentiles: Enter the values of the df, noncentrality parameter, and the cumulative probability; click [x].

Example 17.3.2 When df = 13.0, noncentrality parameter = 2.2, and the cumulative probability = 0.95, the 95th percentile is 26.0113. That is, $P(X \le 26.0113) = 0.95$.

To compute other parameters: Enter the values of one of the parameters, the cumulative probability, and click on the missing parameter.

Example 17.3.3 When df = 13.0, the cumulative probability = 0.95, and $x = 25.0$, the value o the noncentrality parameter is 1.57552.

To compute moments: Enter the values of the df and the noncentrality parameter; click [M].

17.4 Applications

The noncentral chi-square distribution is useful in computing the power of the goodness-of-fit test based on the usual chi-square statistic (see Section 1.4.2)

$$Q = \sum_{i=1}^{k} \frac{(O_i - E_i)^2}{E_i},$$

where O_i is the observed frequency in the ith cell, $E_i = Np_{i0}$ is the expected frequency in the ith cell, p_{i0} is the specified (under the null hypothesis) probability that an observation falls in the ith cell, $i = 1, \cdots, k$, and N = total number of observations. The null hypothesis will be rejected if

$$Q = \sum_{i=1}^{k} \frac{(O_i - E_i)^2}{E_i} > \chi^2_{k-1,\,1-\alpha},$$

where $\chi^2_{k-1,1-\alpha}$ denotes the 100(1 - α)th percentile of a chi-square distribution with df $= k-1$. If the true probability that an observation falls in the ith cell is p_i, $i = 1, \cdots, k$, then Q is approximately distributed as a noncentral chi-square random variable with the noncentrality parameter

$$\delta = N \sum_{i=1}^{k} \frac{(p_i - p_{i0})^2}{p_{i0}},$$

and df $= k - 1$. Thus, an approximate power function is given by

$$P\left(\chi^2_{k-1}(\delta) > \chi^2_{k-1,1-\alpha}\right).$$

The noncentral chi-square distribution is also useful in computing approximate tolerance factors for univariate (see Section 10.6.1) and multivariate (see Section 35.1) normal populations.

17.5 Properties and Results

17.5.1 Properties

1. Let $X_1, \ldots, X_n$ be independent normal random variables with $X_i \sim N(\mu_i, 1)$, $i = 1, 2, ..., n$, and let $\delta = \sum_i^n \mu_i^2$. Then

$$\sum_{i=1}^{n} X_i^2 \sim \chi_n^2(\delta).$$

2. For any real valued function h,

$$E[h(\chi_n^2(\delta))] = E[E(h(\chi_{n+2K}^2)|K)],$$

where K is a Poisson random variable with mean $\delta/2$.

17.5.2 Approximations to Probabilities

Let $a = n + \delta$ and $b = \delta/(n + \delta)$.

1. Let Y be a chi-square random variable with df $= a/(1+b)$. Then

$$P(\chi_n^2(\delta) \le x) \simeq P\left(Y \le \frac{x}{1+b}\right).$$

2. Let Z denote the standard normal random variable. Then

a. $P(\chi_n^2(\delta) \le x) \simeq P\left(Z \le \frac{\left(\frac{x}{a}\right)^{1/3} - \left[1 - \frac{2}{9}\left(\frac{1+b}{a}\right)\right]}{\sqrt{\frac{2}{9}\left(\frac{1+b}{a}\right)}}\right).$

b. $P(\chi_n^2(\delta) \le x) \simeq P\left(Z \le \sqrt{\frac{2x}{1+b}} - \sqrt{\frac{2a}{1+b} - 1}\right).$

17.5.3 Approximations to Percentiles

Let $\chi_{n,p}^2(\delta)$ denote the 100pth percentile of the noncentral chi-square distribution with df $= n$, and noncentrality parameter δ. Define $a = n + \delta$ and $b = \delta/(n + \delta)$

1. Patnaik's (1949) Approximation:

$$\chi^2_{n,p}(\delta) \simeq c\chi^2_{f,p},$$

where $c = 1 + b$, and $\chi^2_{f,p}$ denotes the 100pth percentile of the central chi-square distribution with df $f = a/(1+b)$.

2. Normal Approximations: Let z_p denote the 100pth percentile of the standard normal distribution.

a. $\chi^2_{n,p}(\delta) \simeq \frac{1+b}{2}\left(z_p + \sqrt{\frac{2a}{1+b} - 1}\right)^2.$

b. $\chi^2_{n,p}(\delta) \simeq a\left[z_p\sqrt{\frac{2}{9}\left(\frac{1+b}{a}\right)} - \frac{2}{9}\left(\frac{1+b}{a}\right) + 1\right]^3.$

17.6 Random Number Generation

The following exact method can be used to generate random numbers when the degrees of freedom $n \geq 1$. The following algorithm is based on the additive property of the noncentral chi-square distribution given in Section 17.5.1.

Algorithm 17.6.1

For a given n and δ:
Set u = sqrt(δ)
Generate z_1 from N(u, 1)
Generate y from gamma($(n-1)/2, 2$)
return x = $z_1^2 + y$

x is a random variate from $\chi^2_n(\delta)$ distribution.

17.7 Evaluating the Distribution Function

The following computational method is due to Benton and Krishnamoorthy (2003), and is based on the following infinite series expression for the cdf.

$$P(\chi^2_n(\delta) \leq x) = \sum_{i=0}^{\infty} P(X = i) I_{x/2}(n/2 + i), \tag{17.7.1}$$

where X is a Poisson random variable with mean $\delta/2$, and

$$I_y(a) = \frac{1}{\Gamma(a)} \int_0^y e^{-t} t^{a-1} dt, \quad a > 0,\ y > 0, \tag{17.7.2}$$

is the incomplete gamma function. To compute (17.7.1), evaluate first the kth term, where k is the integer part of $\delta/2$, and then compute the other Poisson probabilities and incomplete gamma functions recursively using forward and backward recursions. To compute Poisson probabilities, use the relations

$$P(X = k+1) = \frac{\delta/2}{k+1} P(X = k), \quad k = 0, 1, \ldots,$$

and

$$P(X = k-1) = \frac{k}{\delta/2} P(x = k), \quad k = 1, 2, \ldots$$

To compute the incomplete gamma function, use the relations

$$I_x(a+1) = I_x(a) - \frac{x^a \exp(-x)}{\Gamma(a+1)}, \tag{17.7.3}$$

and

$$I_x(a-1) = I_x(a) + \frac{x^{a-1} \exp(-x)}{\Gamma(a)}. \tag{17.7.4}$$

Furthermore, the series expansion

$$I_x(a) = \frac{x^a \exp(-x)}{\Gamma(a+1)} \left(1 + \frac{x}{(a+1)} + \frac{x^2}{(a+1)(a+2)} + \cdots \right)$$

can be used to evaluate $I_x(a)$.

When computing the terms using both forward and backward recurrence relations, stop if

$$1 - \sum_{j=k-i}^{k+i} P(X = j)$$

is less than the error tolerance or the number of iterations is greater than a specified integer. While computing using only forward recurrence relation, stop if

$$\left(1 - \sum_{j=0}^{2k+i} P(X = j) \right) I_x(2k + i + 1)$$

is less than the error tolerance or the number of iterations is greater than a specified integer.

The following Fortran function subroutine computes the noncentral chi-square cdf, and is based on the algorithm given in Benton and Krishnamoorthy (2003).

```
Input:
     xx = the value at which the cdf is evaluated, xx > 0
     df = degrees of freedom > 0
     elambda = noncentrality parameter, elambda > 0

Output:
     P(X <= x) = chncdf(x, df, elambda)

cccccccccccccccccccccccccccccccccccccccccccccccccccccccccccccc
      double precision function chncdf(xx, df, elambda)
      implicit double precision (a-h,o-z)
      data one, half, zero/1.0d0, 0.5d0, 0.0d0/
      data maxitr, eps/1000, 1.0d-12/

      chncdf = zero
      if(xx .le. zero) return

      x = half*xx
      del = half*elambda
      k = int(del)
      a = half*df + k

c gamcdf = gamcdf in Section 15.8

      gamkf = gamcdf(x, a)
      gamkb = gamkf

      chncdf = gamkf
      if(del .eq. zero) return

c poipro = Poisson probability

      poikf = poipro(k, del)
      poikb = poikf

c alng(x) = logarithmic gamma function in Section 1.8

      xtermf = dexp((a-one)*dlog(x)-x-alng(a))
      xtermb = xtermf*x/a
      sum = poikf * gamkf
      remain = one - poikf
```

```
1       i = i + 1
          xtermf = xtermf*x/(a+i-one)
          gamkf = gamkf - xtermf
          poikf = poikf * del/(k+i)
          termf = poikf * gamkf
          sum = sum + termf
          error = remain * gamkf
          remain = remain - poikf
c Do forward and backward computations "maxitr" times or until
c convergence
        if (i .gt. k) then
          if(error .le. eps .or. i .gt. maxitr) goto 2
          goto 1
        else
          xtermb = xtermb * (a-i+one)/x
          gamkb = gamkb + xtermb
          poikb = poikb * (k-i+one)/del
          termb = gamkb * poikb
          sum = sum + termb
          remain = remain - poikb
          if(remain .le. eps .or. i .gt. maxitr) goto 2
          goto 1
        end if
2       chncdf = sum
        end
```

Chapter 18

Noncentral F Distribution

18.1 Description

Let $\chi^2_m(\delta)$ be a noncentral chi-square random variable with degrees of freedom (df) $= m$, and noncentrality parameter δ, and χ^2_n be a chi-square random variable with df $= n$. If $\chi^2_m(\delta)$ and χ^2_n are independent, then the distribution of the ratio

$$F_{m,n}(\delta) = \frac{\chi^2_m(\delta)/m}{\chi^2_n/n}$$

is called the noncentral F distribution with the numerator df $= m$, the denominator df $= n$, and the noncentrality parameter δ.

The cumulative distribution function is given by

$$F(x|m,n,\delta) = \sum_{k=0}^{\infty} \frac{\exp(-\frac{\delta}{2})(\frac{\delta}{2})^k}{k!} P\left(F_{m+2k,n} \le \frac{mx}{m+2k}\right),$$
$$m > 0,\ n > 0,\ \delta > 0,$$

where $F_{a,b}$ denotes the central F random variable with the numerator df $= a$, and the denominator df $= b$.

The plots of pdfs of $F_{m,n}(\delta)$ are presented in Figure 18.1 for various values of m, n and δ. It is clear from the plots that the noncentral F distribution is always right skewed.

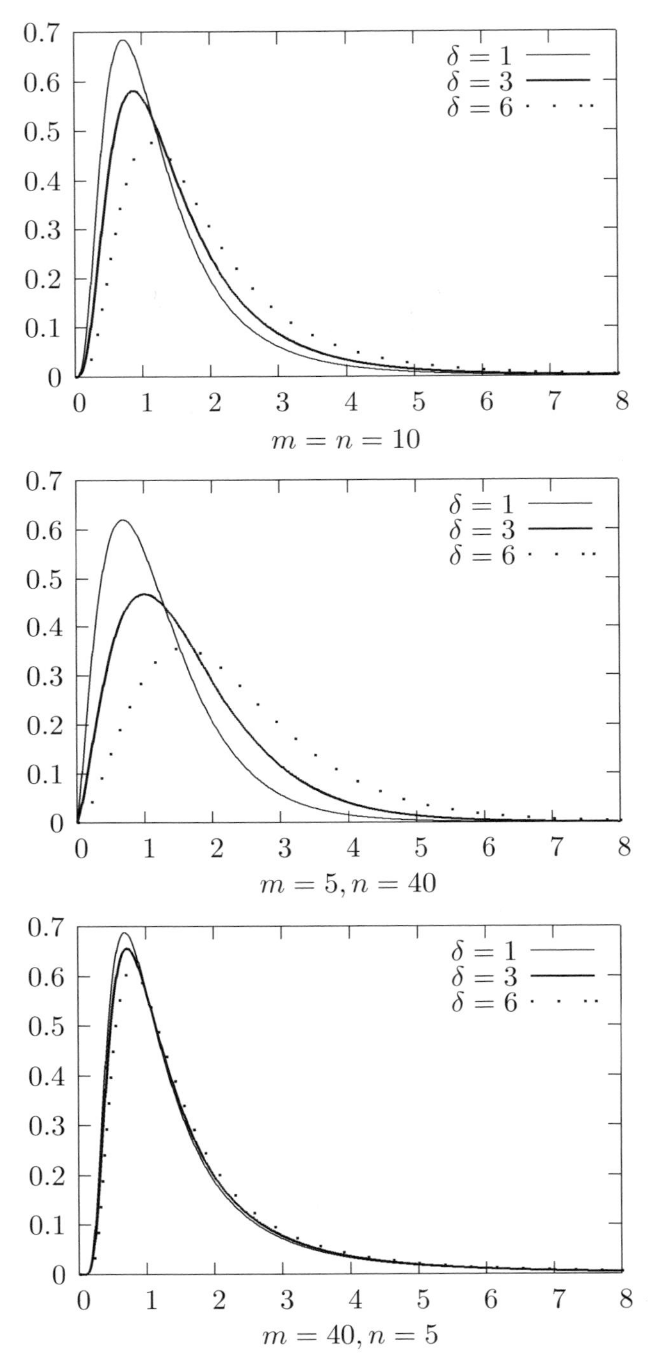

Figure 18.1 Noncentral F pdfs

18.2 Moments

Mean:	$\frac{n(m+\delta)}{m(n-2)}$,	$n > 2$.
Variance:	$\frac{2n^2[(m+\delta)^2+(m+2\delta)(n-2)]}{m^2(n-2)^2(n-4)}$,	$n > 4$.
$E(F_{m,n}^k)$:	$\frac{\Gamma[(n-2k)/2]\,\Gamma[(m+2k)/2]\,n^k}{\Gamma(n/2)m^k} \sum_{j=0}^{k} \binom{k}{j} \frac{(\delta/2)^j}{\Gamma[(m+2j)/2]}$,	$n > 2k$.

18.3 Computing Table Values

The dialog box [StatCalc→Continuous→NC F] computes cumulative probabilities, percentiles, moments and other parameters of an $F_{m,n}(\delta)$ distribution.

To compute probabilities: Enter the values of the numerator df, denominator df, noncentrality parameter, and x; click [P(X <= x)].

Example 18.3.1 When numerator df = 4.0, denominator df = 32.0, noncentrality parameter = 2.2, and $x = 2$, $P(X \leq 2) = 0.702751$ and $P(X > 2) = 0.297249$.

To compute percentiles: Enter the values of the df, noncentrality parameter, and the cumulative probability; click [x].

Example 18.3.2 When numerator df = 4.0, denominator df = 32.0, noncentrality parameter = 2.2, and the cumulative probability = 0.90, the 90th percentile is 3.22243. That is, $P(X \leq 3.22243) = 0.90$.

To compute moments: Enter the values of the numerator df, denominator df and the noncentrality parameter; click [M].

StatCalc also computes one of the degrees of freedoms or the noncentrality parameter for given other values. For example, when numerator df = 5, denominator df = 12, $x = 2$ and $P(X \leq x) = 0.7$, the value of the noncentrality parameter is 2.24162.

18.4 Applications

The noncentral F distribution is useful to compute the powers of a test based on the central F statistic. Examples include analysis of variance and tests based on the Hotelling T^2 statistics. Let us consider the power function of the Hotelling T^2 test for testing about a multivariate normal mean vector.

Let $\boldsymbol{X}_1, \ldots, \boldsymbol{X}_n$ be sample from an m-variate normal population with mean vector $\boldsymbol{\mu}$ and covariance matrix $\boldsymbol{\Sigma}$. Define

$$\bar{\boldsymbol{X}} = \frac{1}{n}\sum_{i=1}^{n} X_i \quad \text{and} \quad S = \frac{1}{n-1}\sum_{i=1}^{n}(X_i - \bar{X})(X_i - \bar{X})'.$$

The Hotelling T^2 statistic for testing $H_0 : \boldsymbol{\mu} = \boldsymbol{\mu}_0$ vs. $H_a : \boldsymbol{\mu} \neq \boldsymbol{\mu}_0$ is given by

$$T^2 = n\left(\bar{\boldsymbol{X}} - \boldsymbol{\mu}_0\right)' S^{-1}\left(\bar{\boldsymbol{X}} - \boldsymbol{\mu}_0\right).$$

Under H_0, $T^2 \sim \frac{(n-1)m}{n-m} F_{m,n-m}$. Under H_a,

$$T^2 \sim \frac{(n-1)m}{n-m} F_{m,n-m}(\delta),$$

where $F_{m,n-m}(\delta)$ denotes the noncentral F random variable with the numerator df $= m$, denominator df $= n - m$, and the noncentrality parameter $\delta = n(\boldsymbol{\mu} - \boldsymbol{\mu}_0)'\boldsymbol{\Sigma}^{-1}(\boldsymbol{\mu} - \boldsymbol{\mu}_0)$ and $\boldsymbol{\mu}$ is true mean vector. The power of the T^2 test is given by

$$P\left(F_{m,n-m}(\delta) > F_{m,n-m,1-\alpha}\right),$$

where $F_{m,n-m,1-\alpha}$ denotes the 100(1 - α)th percentile of the F distribution with the numerator df $= m$ and denominator df $= n - m$.

The noncentral F distribution also arises in multiple use confidence estimation in a multivariate calibration problem. [Mathew and Zha (1996)]

18.5 Properties and Results

18.5.1 Properties

1.

$$\frac{mF_{m,n}(\delta)}{n + mF_{m,n}(\delta)} \sim \text{noncentral beta}\left(\frac{m}{2}, \frac{n}{2}, \delta\right).$$

2. Let $F(x; m, n, \delta)$ denote the cdf of $F_{m,n}(\delta)$. Then

   a. for a fixed m, n, x, $F(x; m, n, \delta)$ is a nonincreasing function of δ;

   b. for a fixed δ, n, x, $F(x; m, n, \delta)$ is a nondecreasing function of m.

18.5.2 Approximations

1. For a large n, $F_{m,n}(\delta)$ is distributed as $\chi^2_m(\delta)/m$.

2. For a large m, $F_{m,n}(\delta)$ is distributed as $(1+\delta/m)\chi^2_n(\delta)$.

3. For large values of m and n,

$$\frac{F_{m,n}(\delta) - \frac{n(m+\delta)}{m(n-2)}}{\frac{n}{m}\left[\frac{2}{(n-2)(n-4)}\left(\frac{(m+\delta)^2}{n-2} + m + 2\delta\right)\right]^{1/2}} \sim N(0,\ 1) \text{ approximately.}$$

4. Let $m^* = \frac{(m+\delta)^2}{m+2\delta}$. Then

$$\frac{m}{m+\delta}F_{m,n}(\delta) \quad \sim \quad F_{m*,n} \quad \text{approximately.}$$

5.

$$\frac{\left(\frac{mF_{m,n}(\delta)}{m+\delta}\right)^{1/3}\left(1-\frac{2}{9n}\right) - \left(1 - \frac{2(m+2\delta)}{9(m+\delta)^2}\right)}{\left[\frac{2(m+2\delta)}{9(m+\delta)^2} + \frac{2}{9n}\left(\frac{mF_{m,n}(\delta)}{m+\delta}\right)^{2/3}\right]^{1/2}} \quad \sim \quad N(0,\ 1) \text{ approximately.}$$

[Abramowitz and Stegun 1965]

18.6 Random Number Generation

The following algorithm is based on the definition of the noncentral F distribution given in Section 18.1.

Algorithm 18.6.1

1. Generate x from the noncentral chi-square distribution with df $= m$ and noncentrality parameter δ (See Section 17.6).

2. Generate y from the central chi-square distribution with df $= n$.

3. return $F = nx/(my)$.

F is a noncentral $F_{m,n}(\delta)$ random number.

18.7 Evaluating the Distribution Function

The following approach is similar to the one for computing the noncentral χ^2 in Section 17.7, and is based on the method for computing the tail probabilities of a noncentral beta distribution given in Chattamvelli and Shanmugham (1997). The distribution function of $F_{m,n}(\delta)$ can be expressed as

$$P(X \le x|m, n, \delta) = \sum_{i=0}^{\infty} \frac{\exp(-\delta/2)(\delta/2)^i}{i!} I_y(m/2 + i,\ n/2), \qquad (18.7.1)$$

where $y = mx/(mx + n)$, and

$$I_y(a,\ b) = \frac{\Gamma(a+b)}{\Gamma(a)\Gamma(b)} \int_0^y t^{a-1}(1-t)^{b-1} dt$$

is the incomplete beta function. Let Z denote the Poisson random variable with mean $\delta/2$. To compute the cdf, compute first the kth term in the series (18.7.1), where k is the integral part of $\delta/2$, and then compute other terms recursively. For Poisson probabilities one can use the forward recurrence relation

$$P(X = k+1|\lambda\ ) = \frac{\lambda}{k+1} p(X = k|\lambda\ ), \quad k = 0, 1, 2, \ldots,$$

and backward recurrence relation

$$P(X = k-1|\lambda\ ) = \frac{k}{\lambda} P(X = k|\lambda\ ), \quad k = 1, 2, \ldots, \qquad (18.7.2)$$

To compute incomplete beta function, use forward recurrence relation

$$I_x(a+1,\ b) = I_x(a,\ b) - \frac{\Gamma(a+b)}{\Gamma(a)\Gamma(b)} x^a (1-x)^b,$$

and backward recurrence relation

$$I_x(a-1,\ b) = I_x(a,\ b) + \frac{\Gamma(a+b-1)}{\Gamma(a)\Gamma(b)} x^{a-1}(1-x)^b. \qquad (18.7.3)$$

While computing the terms using both forward and backward recursions, stop if

$$1 - \sum_{j=k-i}^{k+i} P(X = j)$$

is less than the error tolerance or the number of iterations is greater than a specified integer; otherwise stop if

$$\left(1 - \sum_{j=0}^{2k+i} P(X = j)\right) I_x(m/2 + 2k + i,\ n/2)$$

is less than the error tolerance or the number of iterations is greater than a specified integer.

The following Fortran function subroutine evaluates the cdf of the noncentral F distribution function with numerator df = dfn, denominator df = dfd and the noncentrality parameter "del".

```
Input:
      x = the value at which the cdf is evaluated, x > 0
      dfn = numerator df, dfn > 0
      dfd = denominator df, dfd > 0
      del = noncentrality parameter, del > 0

Output:
      P(X <= x) = cdfncf(x, dfn, dfd, del)

cccccccccccccccccccccccccccccccccccccccccccccccccccccc
      double precision function cdfncf(x, dfn, dfd, del)
      implicit doubleprecision (a-h,o-z)

      data one, half, zero, eps/1.0d0, 0.5d0, 0.0d0, 1.0d-12/

      cdf = zero
      if(x .le. zero) goto 1

      d = half*del
      y = dfn*x/(dfn*x+dfd)
      b = half*dfd
      k = int(d)
      a = half*dfn+k

c betadf(x, a, b) = beta distribution function in Section 16.8

      fkf = betadf(y,a,b)
      cdf = fkf
      if(d .eq. zero) goto 1

c poiprob(k,d) = Poisson pmf given in Section 5.13

      pkf = poiprob(k,d)
      fkb = fkf
```

```
      pkb = pkf

c Logarithmic gamma function alng(x) in Section 1.8 is required

      xtermf = dexp(alng(a+b-one)-alng(a)-alng(b)
     &        +(a-one)*dlog(y)+ b*dlog(one-y))
      xtermb = xtermf*y*(a+b-one)/a

      cdf = fkf*pkf
      sumpois = one - pkf

      if(k .eq. 0) goto 2

      do i = 1, k
         xtermf = xtermf*y*(a+b+(i-one)-one)/(a+i-one)
         fkf = fkf - xtermf
         pkf = pkf * d/(k+i)
         termf = fkf*pkf
         xtermb = xtermb *(a-i+one)/(y*(a+b-i))
         fkb = fkb + xtermb
         pkb = (k-i+one)*pkb/d
         termb = fkb*pkb
         term = termf + termb
         cdf = cdf + term
         sumpois = sumpois-pkf-pkb
         if (sumpois .le. eps) goto 1
      end do

2     xtermf = xtermf*y*(a+b+(i-one)-one)/(a+i-one)
      fkf = fkf - xtermf
      pkf = pkf*d/(k+i)
      termf = fkf*pkf
      cdf = cdf + termf
      sumpois = sumpois-pkf
      if(sumpois <= eps) goto 1
      i = i + 1
      goto 2

1     cdfncf = cdf
      end
```

Chapter 19

Noncentral t Distribution

19.1 Description

Let X be a normal random variable with mean δ and variance 1 and S^2 be a chi-square random variable with degrees of freedom (df) n. If X and S^2 are independent, then the distribution of the ratio $\sqrt{n}X/S$ is called the noncentral t distribution with the degrees of freedom n and the noncentrality parameter δ. The probability density function is given by

$$f(x|n,\delta) = \frac{n^{n/2}\exp(-\delta^2/2)}{\sqrt{\pi}\,\Gamma(n/2)(n+x^2)^{(n+1)/2}}\sum_{i=0}^{\infty}\frac{\Gamma[(n+i+1)/2]}{i!}\left(\frac{x\delta\sqrt{2}}{\sqrt{n+x^2}}\right)^i,$$
$$-\infty < x < \infty, \quad -\infty < \delta < \infty,$$

where $\left(\frac{x\delta\sqrt{2}}{\sqrt{n+x^2}}\right)^0$ should be interpreted as 1 for all values of x and δ, including 0. The above noncentral t random variable is denoted by $t_n(\delta)$.

The noncentral t distribution specializes to Student's t distribution when $\delta = 0$. We also observe from the plots of pdfs in Figure 19.1 that the noncentral t random variable is stochastically increasing with respect to δ. That is, for $\delta_2 > \delta_1$,

$$P(t_n(\delta_2) > x) > P(t_n(\delta_1) > x) \text{ for every } x.$$

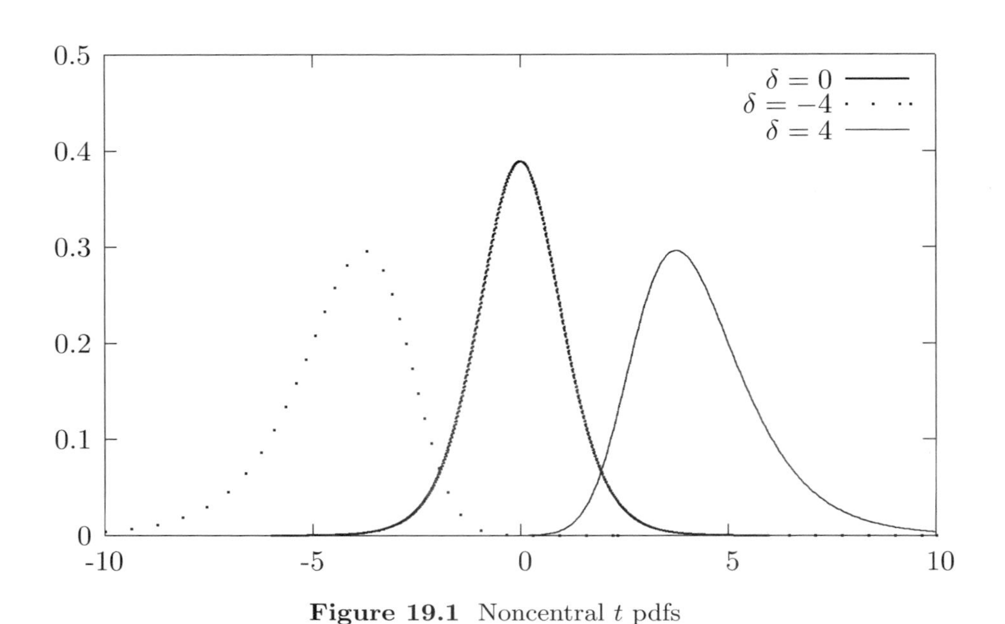

Figure 19.1 Noncentral t pdfs

19.2 Moments

Mean:	$\mu_1 = \frac{\Gamma[(n-1)/2]\sqrt{n/2}}{\Gamma(n/2)}\delta$
Variance:	$\mu_2 = \frac{n}{n-2}(1+\delta^2) - \left(\frac{\Gamma[(n-1)/2]}{\Gamma(n/2)}\right)^2 (n/2)\delta^2$
Moments about the Origin:	$E(X^k) = \frac{\Gamma[(n-k)/2]n^{k/2}}{2^{k/2}\Gamma(n/2)}u_k$, where $u_{2k-1} = \sum_{i=1}^{k} \frac{(2k-1)!\delta^{2i-1}}{(2i-1)!(k-i)!2^{k-i}}$, $k = 1, 2, \ldots$ and $u_{2k} = \sum_{i=0}^{k} \frac{(2k)!\delta^{2i}}{(2i)!(k-i)!2^{k-i}}$, $k = 1, 2, \ldots$ [Bain 1969]
Coefficient of Skewness:	$\frac{\mu_1 \frac{n(2n-3+\delta^2)}{(n-2)(n-3)} - 2\mu_2}{\mu_2^{3/2}}$
Coefficient of Kurtosis:	$\frac{\frac{n^2}{(n-2)(n-4)}(3+6\delta^2+\delta^4) - (\mu_1)^2\left[\frac{n[(n+1)\delta^2+3(3n-5)]}{(n-2)(n-3)} - 3\mu_2\right]}{\mu_2^2}$. [Johnson and Kotz 1970, p. 204]

19.3 Computing Table Values

The dialog box [StatCalc→Continuous→NC t] computes the cdf, percentiles, moments, and noncentrality parameter.

To compute probabilities: Enter the values of the degrees of freedom (df), noncentrality parameter and x; click [P(X <= x)].

Example 19.3.1 When df = 13.0, noncentrality parameter = 2.2 and $x = 2.2$, $P(X \leq 2.2) = 0.483817$ and $P(X > 2.2) = 0.516183$.

To compute percentiles: Enter the values of the df, noncentrality parameter, and the cumulative probability; click [x].

Example 19.3.2 When df = 13.0, noncentrality parameter = 2.2, and the cumulative probability = 0.90, the 90th percentile is 3.87082. That is, $P(X \leq 3.87082) = 0.90$.

To compute other parameters: Enter the values of one of the parameters, the cumulative probability and x. Click on the missing parameter.

Example 19.3.3 When df = 13.0, the cumulative probability = 0.40, and $x = 2$, the value of noncentrality parameter is 2.23209.

To compute moments: Enter the values of the df, and the noncentrality parameter; click [M].

19.4 Applications

The noncentral t distribution arises as a power function of a test if the test procedure is based on a central t distribution. More specifically, powers of the t-test for a normal mean and of the two-sample t-test (Sections 10.4 and 10.5) can be computed using noncentral t distributions. The percentiles of noncentral t distributions are used to compute the one-sided tolerance factors for a normal population (Section 10.6) and tolerance limits for the one-way random effects model (Section 10.6.5). This distribution also arises in multiple-use hypothesis testing about the explanatory variable in calibration problems [Krishnamoorthy, Kulkarni and Mathew (2001), and Benton, Krishnamoorthy and Mathew (2003)].

19.5 Properties and Results

19.5.1 Properties

1. The noncentral distribution $t_n(\delta)$ specializes to the t distribution with df $= n$ when $\delta = 0$.

2. $P(t_n(\delta) \le 0) = P(Z \le -\delta)$, where Z is the standard normal random variable.

3. $P(t_n(\delta) \le t) = P(t_n(-\delta) \ge -t)$.

4. **a.** $P(0 < t_n(\delta) < t) = \sum\limits_{j=0}^{\infty} \frac{\exp(-\delta^2/2)(\delta^2/2)^{j/2}}{\Gamma(j/2+1)} P\left(Y_j \le \frac{t^2}{n+t^2}\right)$,

   b. $P(|t_n(\delta)| < t) = \sum\limits_{j=0}^{\infty} \frac{\exp(-\delta^2/2)(\delta^2/2)^{j}}{j!} P\left(Y_j \le \frac{t^2}{n+t^2}\right)$,
   where Y_j denotes the beta$((j+1)/2, n/2)$ random variable, $j = 1, 2, \ldots$ [Craig 1941 and Guenther 1978].

5.
$$\begin{aligned} P(0 < t_n(\delta) < t) &= \sum_{j=0}^{\infty} \frac{\exp(-\delta^2/2)(\delta^2/2)^j}{j!} P\left(Y_{1j} \le \frac{t^2}{n+t^2}\right) \\ &+ \frac{\delta}{2\sqrt{2}} \sum_{j=0}^{\infty} \frac{\exp(-\delta^2/2)(\delta^2/2)^j}{\Gamma(j+3/2)} P\left(Y_{2j} \le \frac{t^2}{n+t^2}\right), \end{aligned}$$

   where Y_{1j} denotes the beta$((j+1)/2, n/2)$ random variable and Y_{2j} denotes the beta$(j+1, n/2)$ random variable, $j = 1, 2, \ldots$ [Guenther 1978].

6. Relation to the Sample Correlation Coefficient: Let R denote the correlation coefficient of a random sample of $n+2$ observations from a bivariate normal population. Then, letting

$$\rho = \delta\sqrt{2/(2n+1+\delta^2)},$$

   the following function of R,

$$\frac{R}{\sqrt{1-R^2}}\sqrt{\frac{n(2n+1)}{2n+1+\delta^2}} \sim t_n(\delta) \text{ approximately.}$$

   [Harley 1957]

19.5.2 An Approximation

Let $X = t_n(\delta)$. Then

$$Z = \frac{X\left(1 - \frac{1}{4n}\right) - \delta}{\left(1 + \frac{X^2}{2n}\right)^{1/2}} \sim N(0,1) \text{ approximately.}$$

[Abramowitz and Stegun 1965, p 949.]

19.6 Random Number Generation

The following algorithm for generating $t_n(\delta)$ variates is based on the definition given in Section 19.1.

Algorithm 19.6.1

Generate z from N(0, 1)
Set w = z + δ
Generate y from gamma(n/2, 2)
return x = w*sqrt(n)/sqrt(y)

19.7 Evaluating the Distribution Function

The following method is due to Benton and Krishnamoorthy (2003). Letting $x = \frac{t^2}{n+t^2}$, the distribution function can be expressed as

$$\begin{aligned} P(t_n(\delta) \le t) &= \Phi(-\delta) + P(0 < t_n(\delta) \le t) \\ &= \Phi(-\delta) + \frac{1}{2}\sum_{i=0}^{\infty}\left[P_i I_x(i + 1/2,\ n/2) + \frac{\delta}{\sqrt{2}} Q_i I_x(i+1,\ n/2)\right], \end{aligned} \tag{19.7.1}$$

where Φ is the standard normal distribution, $I_x(a, b)$ is the incomplete beta function given by

$$I_x(a,\ b) = \frac{\Gamma(a+b)}{\Gamma(a)\Gamma(b)}\int_0^x y^{a-1}(1-y)^{b-1}dy,$$

$P_i = \exp(-\delta^2/2)(\delta^2/2)^i/i!$ and $Q_i = \exp(-\delta^2/2)(\delta^2/2)^i/\Gamma(i + 3/2)$, $i = 0, 1, 2, ...$

To compute the cdf, first compute the kth term in the series expansion (19.7.1), where k is the integer part of $\delta^2/2$, and then compute the other terms using forward and backward recursions:

$$P_{i+1} = \frac{\delta^2}{2(i+1)}P_i, \qquad P_{i-1} = \frac{2i}{\delta^2}P_i, \qquad Q_{i+1} = \frac{\delta^2}{2i+3}Q_i, \qquad Q_{i-1} = \frac{2i+1}{\delta^2}Q_i$$

$$I_x(a+1,\ b) = I_x(a,\ b) - \frac{\Gamma(a+b)}{\Gamma(a+1)\Gamma(b)}x^a(1-x)^b,$$

and

$$I_x(a-1,\ b) = I_x(a,\ b) + \frac{\Gamma(a+b-1)}{\Gamma(a)\Gamma(b)}x^{a-1}(1-x)^b.$$

Let E_m denote the remainder of the infinite series in (17.7.1) after the mth term. It can be shown that

$$|E_m| \le \frac{1}{2}(1+|\delta|/2)I_x(m+3/2,\ n/2)\left(1-\sum_{i=0}^{m} P_i\right). \tag{19.7.2}$$

[See Lenth 1989 and Benton and Krishnamoorthy 2003]

Forward and backward iterations can be stopped when $1-\sum_{j=k-i}^{k+i} P_j$ is less than the error tolerance or when the number of iterations exceeds a specified integer. Otherwise, forward computation of (19.7.1) can be stopped once the error bound (19.7.2) is less than a specified error tolerance or the number of iterations exceeds a specified integer.

The following Fortran function routine tnd(t, df, delta) computes the cdf of a noncentral t distribution. This program is based on the algorithm given in Benton and Krishnamoorthy (2003).

```
cccccccccccccccccccccccccccccccccccccccccccccccccccccccccccccccc
      double precision function tnd(t, df, delta)
      implicit double precision (a-h, o-z)
      logical indx
      data zero, half, one /0.0d0, 0.5d0, 1.0d0/
      data error, maxitr/1.0d-12, 1000/
c
      if (t .lt. zero) then
      x = -t
      del = -delta
      indx = .true.
      else
```

```
      x = t
      del = delta
      indx = .false.
      end if

c gaudf(x) is the normal cdf in Section 10.10

      ans = gaudf(-del)
      if( x .eq. zero) then
      tnd = ans
      return
      end if
c
      y = x*x/(df+x*x)
      dels = half*del*del
      k = int(dels)
      a = k+half
      c = k+one
      b = half*df

c alng(x) is the logarithmic gamma function in Section 1.8

      pkf = dexp(-dels+k*dlog(dels)-alng(k+one))
      pkb = pkf
      qkf = dexp(-dels+k*dlog(dels)-alng(k+one+half))
      qkb = qkf

c betadf(y, a, b) is the beta cdf in Section 16.6

      pbetaf = betadf(y, a, b)
      pbetab = pbetaf
      qbetaf = betadf(y, c, b)
      qbetab = qbetaf
      pgamf = dexp(alng(a+b-one)-alng(a)-alng(b)+(a-one)*dlog(y)
     +        + b*dlog(one-y))
      pgamb = pgamf*y*(a+b-one)/a
      qgamf = dexp(alng(c+b-one)-alng(c)-alng(b)+(c-one)*dlog(y)
     +        + b*dlog(one-y))
      qgamb = qgamf*y*(c+b-one)/c
c
      rempois = one - pkf
```

```
      delosq2 = del/1.4142135623731d0
      sum = pkf*pbetaf+delosq2*qkf*qbetaf
      cons = half*(one + half*abs(delta))
      i = 0
1     i = i + 1
      pgamf = pgamf*y*(a+b+i-2.0)/(a+i-one)
      pbetaf = pbetaf - pgamf
      pkf = pkf*dels/(k+i)
      ptermf = pkf*pbetaf
      qgamf = qgamf*y*(c+b+i-2.0)/(c+i-one)
      qbetaf = qbetaf - qgamf
      qkf = qkf*dels/(k+i-one+1.5d0)
      qtermf = qkf*qbetaf
      term = ptermf + delosq2*qtermf
      sum = sum + term
      error = rempois*cons*pbetaf
      rempois = rempois - pkf

c  Do forward and backward computations k times or until convergen

      if (i. gt. k) then
         if(error .le. error .or. i .gt. maxitr) goto 2
         goto 1
      else
         pgamb = pgamb*(a-i+one)/(y*(a+b-i))
         pbetab = pbetab + pgamb
         pkb = (k-i+one)*pkb/dels
         ptermb = pkb*pbetab
         qgamb = qgamb*(c-i+one)/(y*(c+b-i))
         qbetab = qbetab + qgamb
         qkb = (k-i+one+half)*qkb/dels
         qtermb = qkb*qbetab
         term =  ptermb + delosq2*qtermb
         sum = sum + term
         rempois = rempois - pkb
         if (rempois .le. error .or. i .ge. maxitr) goto 2
         goto 1
      end if
2     tnd = half*sum + ans
      if(indx) tnd = one - tnd
      end
```

Chapter 20

Laplace Distribution

20.1 Description

The distribution with the probability density function

$$f(x|a,b) = \frac{1}{2b}\exp\left[-\frac{|x-a|}{b}\right], \quad -\infty < x < \infty,\ -\infty < a < \infty,\ b > 0, \tag{20.1.1}$$

where a is the location parameter and b is the scale parameter, is called the Laplace(a, b) distribution.

The cumulative distribution function is given by

$$F(x|a,b) = \begin{cases} 1 - \frac{1}{2}\exp\left[\frac{a-x}{b}\right] & \text{for } x \geq a, \\ \frac{1}{2}\exp\left[\frac{x-a}{b}\right] & \text{for } x < a. \end{cases} \tag{20.1.2}$$

The Laplace distribution is also referred to as the *double exponential* distribution. For any given probability p, the inverse distribution is given by

$$F^{-1}(p|a,b) = \begin{cases} a + b\ln(2p) & \text{for } 0 < p \leq 0.5, \\ a - b\ln(2(1-p)) & \text{for } 0.5 < p < 1. \end{cases} \tag{20.1.3}$$

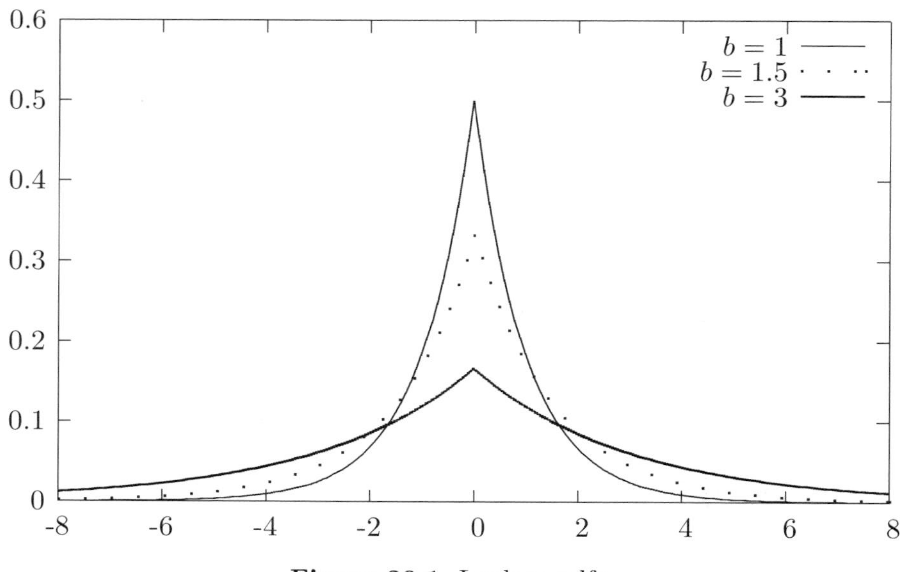

Figure 20.1 Laplace pdfs

20.2 Moments

Mean:	a
Median:	a
Mode:	a
Variance:	$2b^2$
Mean Deviation:	b
Coefficient of Variation:	$\frac{b\sqrt{2}}{a}$
Coefficient of Skewness:	0
Coefficient of Kurtosis:	6
Moments about the Mean:	$E(X-a)^k = \begin{cases} 0 & \text{for } k = 1, 3, 5, \ldots \\ k!b^k & \text{for } k = 2, 4, 6, \ldots \end{cases}$

20.3 Computing Table Values

For given values of a and b, the dialog box [*StatCalc*→Continuous→Laplace] computes the cdf, percentiles, moments, and other parameters of the Laplace(a, b) distribution.

To compute probabilities: Enter the values of the parameters a, b, and the value of x; click [P(X <= x)].

Example 20.3.1 When $a = 3$, $b = 4$, and $x = 4.5$,

$$P(X \leq 4.5) = 0.656355 \quad \text{and} \quad P(X > 4.5) = 0.343645.$$

To compute percentiles: Enter the values of a, b, and the cumulative probability; click [x].

Example 20.3.2 When $a = 3$, $b = 4$, and the cumulative probability $= 0.95$, the 95th percentile is 12.2103. That is, $P(X \leq 12.2103) = 0.95$.

To compute parameters: Enter value of one of the parameters, cumulative probability, and x; click on the missing parameter.

Example 20.3.3 When $a = 3$, cumulative probability $= 0.7$, and $x = 3.2$, the value of b is 0.391523.

To compute moments: Enter the values of a and b and click [M].

20.4 Inferences

Let $X_1, \ldots, X_n$ be a sample of independent observations from a Laplace distribution with the pdf (20.1.1). Let $X_{(1)} < X_{(2)} < \ldots < X_{(n)}$ be the order statistics based on the sample.

20.4.1 Maximum Likelihood Estimators

If the sample size n is odd, then the sample median $\widehat{a} = X_{((n+1)/2)}$ is the MLE of a. If n is even, then the MLE of a is any number between $X_{(n/2)}$ and $X_{(n/2+1)}$. The MLE of b is given by

$$\widehat{b} = \frac{1}{n}\sum_{i=1}^{n}|X_i - \widehat{a}| \ \text{(if } a \text{ is unknown)} \quad \text{and} \quad \widehat{b} = \frac{1}{n}\sum_{i=1}^{n}|X_i - a| \ \text{(if } a \text{ is known).}$$

20.4.2 Interval Estimation

If a is known, then a $1-\alpha$ confidence interval for b is given by

$$\left(\frac{2\sum\limits_{i=1}^{n}|X_i-a|}{\chi^2_{2n,1-\alpha/2}}, \frac{2\sum\limits_{i=1}^{n}|X_i-a|}{\chi^2_{2n,\alpha/2}}\right).$$

20.5 Applications

Because the distribution of differences between two independent exponential variates with mean b is Laplace (0, b), a Laplace distribution can be used to model the difference between the waiting times of two events generated by independent random processes. The Laplace distribution can also be used to describe breaking strength data. Korteoja et al. (1998) studied tensile strength distributions of four paper samples and concluded that among extreme value, Weibull and Laplace distributions, a Laplace distribution fits the data best. Sahli et al. (1997) proposed a one-sided acceptance sampling by variables when the underlying distribution is Laplace. In the following we see an example where the differences in flood stages are modeled by a Laplace distribution.

Example 20.5.1 The data in Table 20.1 represent the differences in flood stages for two stations on the Fox River in Wisconsin for 33 different years. The data were first considered by Gumbel and Mustafi (1967), and later Bain and Engelhardt (1973) justified the Laplace distribution for modeling the data. Kappenman (1977) used the data for constructing one-sided tolerance limits.

To fit a Laplace distribution for the observed differences of flood stages, we estimate

$$\widehat{a} = 10.13 \quad \text{and} \quad \widehat{b} = 3.36$$

by the maximum likelihood estimates (see Section 20.4.1). Using these estimates, the population quantiles are estimated as described in Section 1.4.1. For example, to find the population quantile corresponding to the sample quantile 1.96, select [Continuous→Laplace] from *StatCalc*, enter 10.13 for a, 3.36 for b and 0.045 for [P(X <= x)]; click on [x] to get 2.04.

The Q-Q plot of the observed differences and the Laplace(10.13, 3.36) quantiles is given in Figure 20.2. The Q-Q plot shows that the sample quantiles (the observed differences) and the population quantiles are in good agreement. Thus, we conclude that the Laplace(10.13, 3.36) distribution adequately fits the data on flood stage differences.

Table 20.1 Differences in Flood Stages

j	Observed Differences	$\frac{j-0.5}{33}$	Population Quantiles	j	Observed Differences	$\frac{j-0.5}{33}$	Population Quantiles
1	1.96	–	–	18	10.24	0.530	10.34
2	1.96	0.045	2.04	19	10.25	0.561	10.56
3	3.60	0.076	3.80	20	10.43	0.591	10.80
4	3.80	0.106	4.92	21	11.45	0.621	11.06
5	4.79	0.136	5.76	22	11.48	0.652	11.34
6	5.66	0.167	6.44	23	11.75	0.682	11.65
7	5.76	0.197	7.00	24	11.81	0.712	11.99
8	5.78	0.227	7.48	25	12.34	0.742	12.36
9	6.27	0.258	7.90	26	12.78	0.773	12.78
10	6.30	0.288	8.27	27	13.06	0.803	13.26
11	6.76	0.318	8.61	28	13.29	0.833	13.82
12	7.65	0.348	8.92	29	13.98	0.864	14.50
13	7.84	0.379	9.20	30	14.18	0.894	15.34
14	7.99	0.409	9.46	31	14.40	0.924	16.47
15	8.51	0.439	9.70	32	16.22	0.955	18.19
16	9.18	0.470	9.92	33	17.06	0.985	21.88
17	10.13	0.500	10.13				

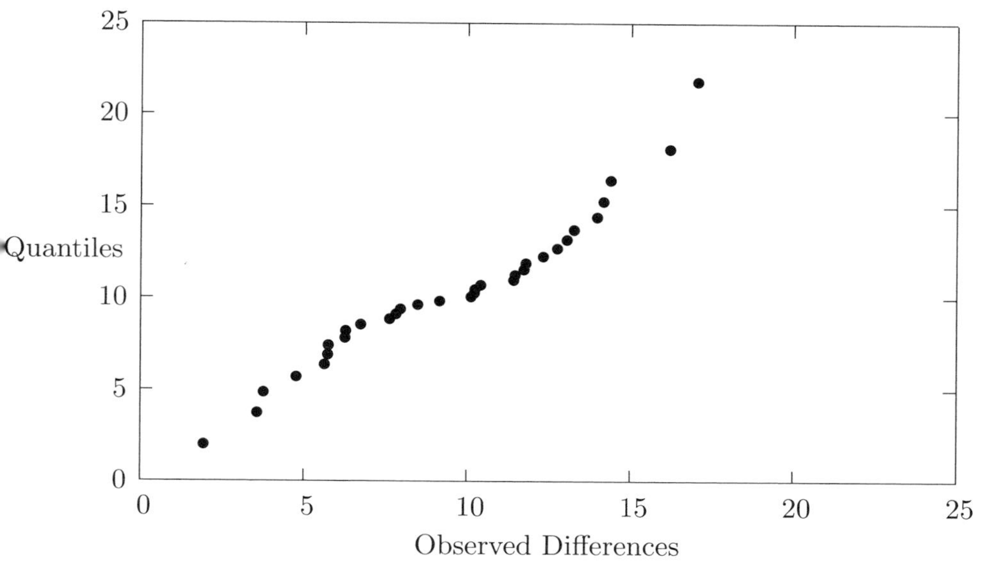

Figure 20.2 Q-Q Plot of Differences in Flood Stages Data

The fitted distribution can be used to estimate the probabilities. For example, the percentage of differences in flood stages exceed 12.4 is estimated by

$$P(X > 12.4|a = 10.13, b = 3.36) = 0.267631.$$

That is, about 27% of differences in flood stages exceed 12.4.

20.6 Relation to Other Distributions

1. Exponential: If X follows a Laplace(a, b) distribution, then $|X - a|/b$ follows an exponential distribution with mean 1. That is, if $Y = |X - a|/b$, then the pdf of Y is $\exp(-y)$, $y > 0$.

2. Chi-square: $|X - a|$ is distributed as $(b/2)\chi_2^2$.

3. Chi-square: If $X_1, \ldots, X_n$ are independent Laplace(a, b) random variables, then

$$\frac{2}{b}\sum_{i=1}^{n}|X_i - a| \sim \chi_{2n}^2.$$

4. F Distribution: If X_1 and X_2 are independent Laplace(a, b) random variables, then

$$\frac{|X_1 - a|}{|X_2 - a|} \sim F_{2,2}.$$

5. Normal: If Z_1, Z_2, Z_3 and Z_4 are independent standard normal random variables, then

$$Z_1 Z_2 - Z_3 Z_4 \sim \text{Laplace}(0,\ 2).$$

6. Exponential: If Y_1 and Y_2 are independent exponential random variables with mean b, then

$$Y_1 - Y_2 \sim \text{Laplace}(0,\ b).$$

7. Uniform: If U_1 and U_2 are uniform(0,1) random variables, then

$$\ln(U_1/U_2) \sim \text{Laplace}(0,\ 1).$$

20.7 Random Number Generation

Algorithm 20.7.1

For a given a and b:
Generate u from uniform(0, 1)
If $u \geq 0.5$, return $x = a - b * \ln(2 * (1 - u))$
else return $x = a + b * \ln(2 * u)$

x is a pseudo random number from the Laplace(a, b) distribution.

Chapter 21

Logistic Distribution

21.1 Description

The probability density function of a logistic distribution with the location parameter a and scale parameter b is given by

$$f(x|a,b) = \frac{1}{b}\frac{\exp\left\{-\left(\frac{x-a}{b}\right)\right\}}{\left[1+\exp\left\{-\left(\frac{x-a}{b}\right)\right\}\right]^2}, \qquad -\infty < x < \infty, \quad -\infty < a < \infty, \quad b > 0. \tag{21.1.1}$$

The cumulative distribution function is given by

$$F(x|a,b) = \left[1+\exp\left\{-\left(\frac{x-a}{b}\right)\right\}\right]^{-1}. \tag{21.1.2}$$

For $0 < p < 1$, the inverse distribution function is given by

$$F^{-1}(p|a,b) = a + b\ln[p/(1-p)]. \tag{21.1.3}$$

The logistic distribution is symmetric about the location parameter a (see Figure 21.1), and it can be used as a substitute for a normal distribution.

21.2 Moments

Mean:	a
Variance:	$\frac{b^2\pi^2}{3}$
Mode:	a
Median:	a
Mean Deviation:	$2b\ln(2)$
Coefficient of Variation:	$\frac{b\pi}{a\sqrt{3}}$
Coefficient of Skewness:	0
Coefficient of Kurtosis:	4.2
Moment Generating Function:	$E(e^{tY}) = \pi\text{cosec}(t\pi)$, where $Y = (X - a)/b$.
Inverse Distribution Function:	$a + b\ln[p/(1-p)]$
Survival Function:	$\frac{1}{1+\exp[(x-a)/b]}$
Inverse Survival Function:	$a + b\ \ln\{(1-p)/p\}$
Hazard Rate:	$\frac{1}{b[1+\exp[-(x-a)/b]]}$
Hazard Function:	$\ln\{1 + \exp[(x-a)/b]\}$

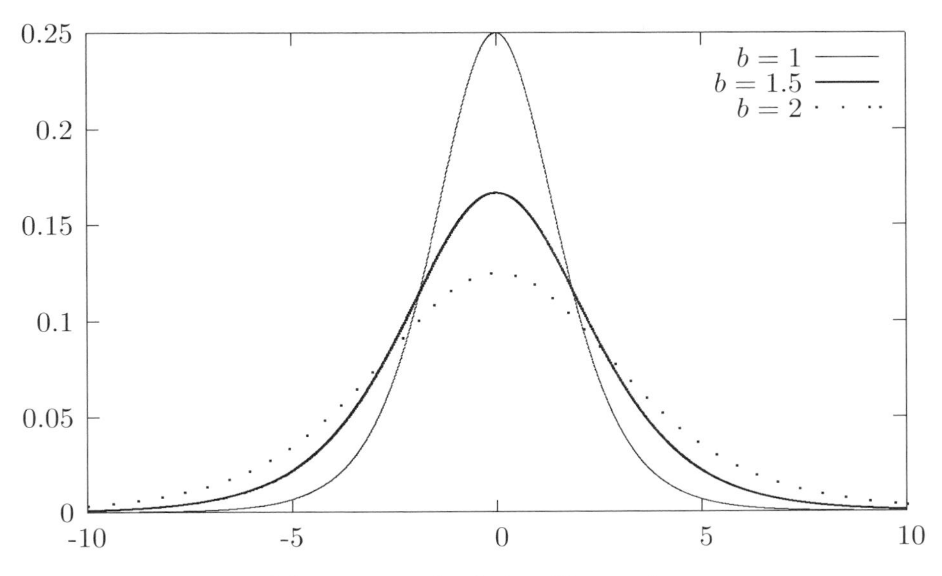

Figure 21.1 Logistic pdfs; $a = 0$

21.3 Computing Table Values

For given values of a and b, the dialog box [StatCalc→Continuous→Logistic] computes the cdf, percentiles and moments of a Logistic(a, b) distribution.

To compute probabilities: Enter the values of the parameters a, b, and the value of x; click [P(X <= x)].

Example 21.3.1 When $a = 2$, $b = 3$, and the observed value $x = 1.3$, $P(X \leq 1.3) = 0.44193$ and $P(X > 1.3) = 0.55807$.

To compute percentiles: Enter the values a, b, and the cumulative probability; click [x].

Example 21.3.2 When $a = 2$, $b = 3$, and the cumulative probability = 0.25, the 25th percentile is -1.29584. That is, $P(X \leq -1.29584) = 0.25$.

To compute other parameters: Enter the values of one of the parameters, cumulative probability and x; click on the missing parameter.

Example 21.3.3 When $b = 3$, cumulative probability = 0.25 and $x = 2$, the value of a is 5.29584.

To compute moments: Enter the values of a and b and click [M].

21.4 Maximum Likelihood Estimators

Let $X_1, \ldots, X_n$ be a sample of independent observations from a logistic distribution with parameters a and b. Explicit expressions for the MLEs of a and b are not available. Likelihood equations can be solved only numerically, and they are

$$\sum_{i=1}^{n} \left[1 + \exp\left(\frac{X_i - a}{b}\right)\right]^{-1} = \frac{n}{2}$$

$$\sum_{i=1}^{n} \left(\frac{X_i - a}{b}\right) \frac{1 - \exp[(X_i - a)/b]}{1 + \exp[(X_i - a)/b]} = n. \qquad (21.4.1)$$

The sample mean and standard deviation can be used to estimate a and b. Specifically,

$$\widehat{a} = \frac{1}{n}\sum_{i=1}^{n} X_i \quad \text{and} \quad \widehat{b} = \frac{\sqrt{3}}{\pi}\sqrt{\frac{1}{n-1}\sum_{i=1}^{n}(X_i - \bar{X})^2}.$$

(See the formula for variance.) These estimators may be used as initial values to solve the equations in (21.4.1) numerically for a and b.

21.5 Applications

The logistic distribution can be used as a substitute for a normal distribution. It is also used to analyze data related to stocks. Braselton et. al. (1999) considered the day-to-day percent changes of the daily closing values of the S&P 500 index from January 1, 1926 through June 11, 1993. These authors found that a logistic distribution provided the best fit for the data even though the lognormal distribution has been used traditionally to model these daily changes. An application of the logistic distribution in nuclear-medicine is given in Prince et. al. (1988). de Visser and van den Berg (1998) studied the size grade distribution of onions using a logistic distribution. The logistic distribution is also used to predict the soil-water retention based on the particle-size distribution of Swedish soil (Rajkai et. al. 1996). Scerri and Farrugia (1996) compared the logistic and Weibull distributions for modeling wind speed data. Applicability of a logistic distribution to study citrus rust mite damage on oranges is given in Yang et. al. (1995).

21.6 Properties and Results

1. If X is a Logistic(a, b) random variable, then $(X - a)/b \sim$ Logistic$(0, 1)$.
2. If u follows a uniform(0, 1) distribution, then $a + b[\ln(u)$ - $\ln(1 - u)] \sim$ Logistic(a, b).
3. If Y is a standard exponential random variable, then
$$-\ln\left[\frac{e^{-y}}{1 - e^{-y}}\right] \sim \text{Logistic}(0,\ 1).$$
4. If Y_1 and Y_2 are independent standard exponential random variables, then
$$-\ln\left(\frac{Y_1}{Y_2}\right) \sim \text{Logistic}(0,\ 1).$$

For more results and properties, see Balakrishnan (1991).

21.7 Random Number Generation

Algorithm 21.7.1

For a given a and b:
Generate u from uniform(0, 1)
return $x = a + b * (\ln(u) - \ln(1 - u))$

x is a pseudo random number from the Logistic(a, b) distribution.

Chapter 22

Lognormal Distribution

22.1 Description

A positive random variable X is lognormally distributed if $\ln(X)$ is normally distributed. The pdf of X is given by

$$f(x|\mu,\sigma) = \frac{1}{\sqrt{2\pi}x\sigma}\exp\left[-\frac{(\ln x-\mu)^2}{2\sigma^2}\right], \quad x>0,\ \sigma>0,\ -\infty<\mu<\infty. \tag{22.1.1}$$

Note that if $Y = \ln(X)$, and Y follows a normal distribution with mean μ and standard deviation σ, then the distribution of X is called lognormal. Since X is actually an antilogarithmic function of a normal random variable, some authors refer to this distribution as antilognormal. We denote this distribution by lognormal(μ, σ^2).

The cdf of a lognormal(μ, σ^2) distribution is given by

$$\begin{aligned} F(x|\mu,\sigma) &= P(X \le x|\mu,\sigma) \\ &= P(\ln X \le \ln x|\mu,\sigma) \\ &= P\left(Z \le \frac{\ln x-\mu}{\sigma}\right) \\ &= \Phi\left(\frac{\ln x-\mu}{\sigma}\right), \end{aligned} \tag{22.1.2}$$

where Φ is the standard normal distribution function.

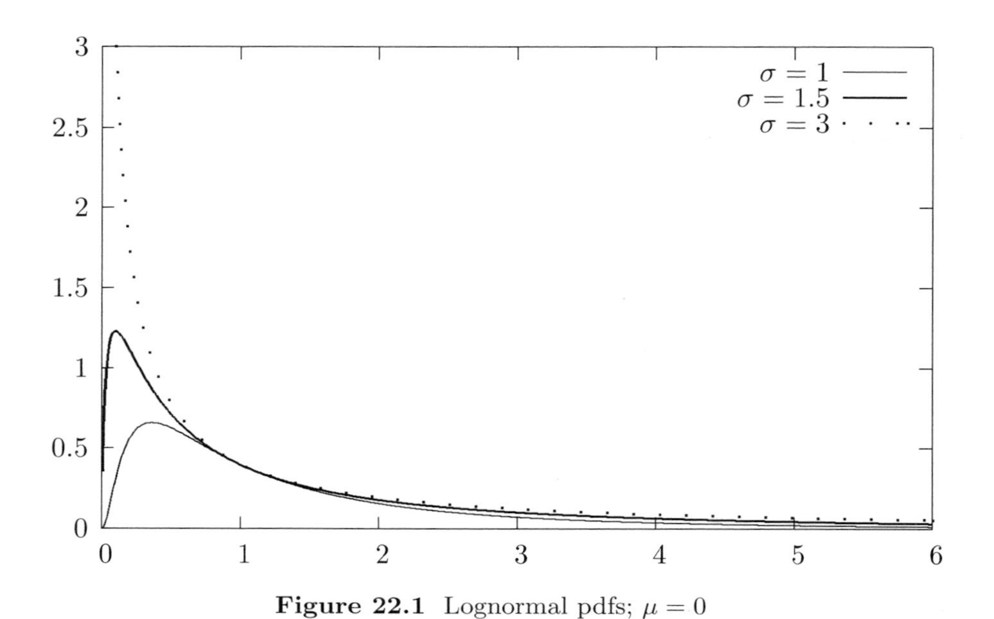

Figure 22.1 Lognormal pdfs; $\mu = 0$

22.2 Moments

Mean:	$\exp[\mu + \sigma^2/2]$
Variance:	$\exp(\sigma^2)[\exp(\sigma^2) - 1]\exp(2\mu)$
Mode:	$\exp[\mu - \sigma^2]$
Median:	$\exp(\mu)$
Coefficient of Variation:	$\sqrt{[\exp(\sigma^2) - 1]}$
Coefficient of Skewness:	$[\exp(\sigma^2) + 2]\sqrt{[\exp(\sigma^2) - 1]}$
Coefficient of Kurtosis:	$\exp(4\sigma^2) + 2\exp(3\sigma^2) + 3\exp(2\sigma^2) - 3$
Moments about the Origin:	$\exp\left[k\mu + k^2\sigma^2/2\right]$
Moments about the Mean:	$\exp\left[k(\mu + \sigma^2/2)\right] \sum_{i=0}^{k} (-1)^i \binom{k}{i} \exp\left[\frac{\sigma^2(k-i)(k-i-1)}{2}\right]$. [Johnson et. al. (1994, p. 212)]

22.3 Computing Table Values

The dialog box [StatCalc→Continuous →Lognormal] computes the cdf, percentiles and moments of a lognormal(μ, σ^2) distribution. This dialog box also computes the following.

1. Confidence Interval and the p-value of a Test about a Lognormal Mean [Section 22.5].

2. Confidence Interval and the p-value of a Test for the Difference Between Two Lognormal Means [Section 22.6].

3. Confidence Interval for the Ratio of Two Lognormal Means [Section 22.7].

To compute probabilities: Enter the values of the parameters μ, σ, and the observed value x; click [P(X <= x)].

Example 22.3.1 When $\mu = 1$, $\sigma = 2$, and $x = 2.3$, $P(X \leq 2.3) = 0.466709$ and $P(X > 2.3) = 0.533291$.

To compute percentiles: Enter the values of μ, σ, and the cumulative probability P(X <= x); click on [x].

Example 22.3.2 When $\mu = 1$, $\sigma = 2$, and the cumulative probability P(X <= x) = 0.95, the 95th percentile is 72.9451. That is, $P(X \leq 72.9451) = 0.95$.

To compute μ: Enter the values of σ, x, and the cumulative probability P(X <= x); click on [U].

Example 22.3.3 When $x = 2.3$, $\sigma = 2$, and the cumulative probability P(X <= x) = 0.9, the value of μ is -1.73019.

To compute σ: Enter the values of x, μ and the cumulative probability P(X <= x); click on [S].

Example 22.3.4 When $x = 3$, $\mu = 2$, and the cumulative probability P(X <= x) = 0.1, the value of σ is 0.703357.

To compute moments: Enter the values of μ and σ and click [M].

22.4 Maximum Likelihood Estimators

Let $X_1, \ldots, X_n$ be a sample of independent observations from a lognormal(μ, σ) distribution. Let $Y_i = \ln(X_i)$, $i = 1, \ldots, n$. Then

$$\widehat{\mu} = \bar{Y} = \frac{1}{n}\sum_{i=1}^{n} Y_i \quad \text{and} \quad \widehat{\sigma} = \sqrt{\frac{1}{n}\sum_{i=1}^{n}(Y_i - \bar{Y})^2}$$

are the MLEs of μ and σ, respectively.

22.5 Confidence Interval and Test for the Mean

Let $X_1, ..., X_n$ be a sample from a lognormal(μ, σ). Let $Y_i = \ln(X_i)$, $i = 1, \ldots, n$. Let

$$\bar{Y} = \frac{1}{n}\sum_{i=1}^{n} Y_i \quad \text{and} \quad S^2 = \frac{1}{n-1}\sum_{i=1}^{n}\left(Y_i - \bar{Y}\right)^2.$$

Recall that the mean of a lognormal(μ, σ) distribution is given by $\exp(\eta)$, where $\eta = \mu + \sigma^2/2$. Since the lognormal mean is a one-one function of η, it is enough to estimate or test about η. For example, if L is a 95% lower limit for η, then $\exp(L)$ is a 95% lower limit for $\exp(\eta)$. The inferential procedures are based on the *generalized variable* approach given in Krishnamoorthy and Mathew (2003). For given observed values $\bar{y}$ of $\bar{Y}$ and s of S, the following algorithm can be used to compute interval estimates and p-values for hypothesis testing about the mean.

Algorithm 22.5.1

For $j = 1, m$
Generate $Z \sim N(0,1)$ and $U^2 \sim \chi^2_{n-1}$
Set $T_j = \bar{y} - \sqrt{\frac{n-1}{n}}\frac{Zs}{U} + \frac{1}{2}\frac{(n-1)s^2}{U^2}$
(end loop)

The percentiles of the T_j's generated above can be used to find confidence intervals for η. Let T_p denote the 100pth percentile of the T_j's. Then, $(T_{.025}, T_{.975})$ is a 95% confidence interval for η. Furthermore, $(\exp(T_{.025}), \exp(T_{.975}))$ is a 95% confidence interval for the lognormal mean $\exp(\eta)$. A 95% lower limit for $\exp(\eta)$ is given by $\exp(T_{0.05})$.

Suppose we are interested in testing

$$H_0 : \exp(\eta) \leq c \quad \text{vs} \quad H_a : \exp(\eta) > c,$$

where c is a specified number. Then the p-value based on the generalized variable approach is given by $P(\exp(T) < c) = P(T < \ln(c))$, and it can be estimated by the proportion of the T_i's less than $\ln(c)$. Note that the p-value is given by $P(T < \ln(c))$, because, for fixed $(\bar{y}, s)$, the generalized variable is stochastically decreasing with respect to η.

For a given sample size, mean, and standard deviation of the logged data, *StatCalc* computes confidence intervals and the p-values for testing about a lognormal mean using Algorithm 22.5.1 with m = 1,000,000.

Illustrative Examples

Example 22.5.1 Suppose that a sample of 15 observations from a lognormal(μ, σ) distribution produced the mean of logged observations $\bar{y} = 1.2$ and the standard deviation of logged observations $s = 1.5$. It is desired to find a 95% confidence interval for the lognormal mean $\exp(\mu + \sigma^2/2)$. To compute a 95% confidence interval using *StatCalc*, select [StatCalc→Continuous→Lognormal→CI and Test for Mean], enter 15 for the sample size, 1.2 for [Mean of ln(x)], 1.5 for [Std Dev of ln(x)], 0.95 for the confidence level, and click [2] to get (4.37, 70.34).

To find one-sided confidence limits, click [1] to get 5.38 and 52.62. This means that the true mean is greater than 5.38 with 95% confidence. Furthermore, the true mean is less than 52.62 with 95% confidence.

Suppose we want to test

$$H_0 : \exp(\eta) \leq 4.85 \quad \text{vs} \quad H_a : \exp(\eta) > 4.85.$$

To find the p-value, enter 4.85 for [H0: M = M0] and click [p-values for] to get 0.045. Thus, at 5% level, we can reject the null hypothesis and conclude that the true mean is greater than 4.85.

22.6 Inferences for the Difference between Two Means

Suppose that we have a sample of n_i observations from a lognormal(μ, σ^2) population, $i = 1, 2$. Let $\bar{y}_i$ and s_i denote, respectively, the mean and standard deviation of the logged measurements in the ith sample, $i = 1, 2$. For a given $(n_1, \bar{y}_1, s_1, n_2, \bar{y}_2, s_2)$, *StatCalc* computes confidence intervals and p-values for hypothesis testing about the difference between two lognormal means $\exp(\eta_1) - \exp(\eta_2)$, where $\eta_i = \mu_i + \sigma_i^2/2$, $i = 1, 2$. *StatCalc* uses the following Monte Carlo method:

Algorithm 22.6.1

For $j = 1, m$
Generate independent random numbers Z_1, Z_2, U_1^2 and U_2^2
such that $Z_i \sim N(0,1)$ and $U_i^2 \sim \chi^2_{n_i - 1}$, $i = 1, 2$.
Set
$G_i = \bar{y}_i - \sqrt{\frac{n_i - 1}{n_i}} \frac{Z_i s_i}{U_i} + \frac{1}{2}\frac{(n_i-1)s_i^2}{U_i^2}, i = 1, 2.$
$T_j = \exp(G_1) - \exp(G_2)$
(end loop)

The percentiles of the T_j's generated above can be used to construct confidence intervals for $\exp(\eta_1) - \exp(\eta_2)$. Let T_p denote the 100pth percentile of the T_j's. Then, $(T_{.025}, T_{.975})$ is a 95% confidence interval for $\exp(\eta_1) - \exp(\eta_2)$; $T_{.05}$ is a 95% lower limit for $\exp(\eta_1) - \exp(\eta_2)$.

Suppose we are interested in testing

$$H_0 : \exp(\eta_1) - \exp(\eta_2) \le 0 \quad \text{vs.} \quad H_a : \exp(\eta_1) - \exp(\eta_2) > 0.$$

Then, an estimate of the p-value based on the generalized variable approach is the proportion of the T_j's that are less than 0.

For given sample sizes, sample means, and standard deviations of the logged data, *StatCalc* computes the confidence intervals and the p-values for testing about the difference between two lognormal means using Algorithm 22.6.1 with m = 100,000.

Illustrative Examples

Example 22.6.1 The data for this example are taken from the web site http://lib.stat.cmu.edu/DASL/. An oil refinery conducted a series of 31 daily measurements of the carbon monoxide levels arising from one of their stacks. The measurements were submitted as evidence for establishing a baseline to the Bay Area Air Quality Management District (BAAQMD). BAAQMD personnel also made nine independent measurements of the carbon monoxide concentration from the same stack. The data are given below:

Carbon Monoxide Measurements by the Refinery (in ppm):
45, 30, 38, 42, 63, 43, 102, 86, 99, 63, 58, 34, 37, 55, 58, 153, 75 58, 36, 59, 43, 102, 52, 30, 21, 40, 141, 85, 161, 86, 161, 86, 71
Carbon Monoxide Measurements by the BAAQMD (in ppm):
12.5, 20, 4, 20, 25, 170, 15, 20, 15

The assumption of lognormality is tenable. The hypotheses to be tested are

$$H_0 : \exp(\eta_1) \leq \exp(\eta_2) \quad \text{vs.} \quad H_a : \exp(\eta_1) > \exp(\eta_2),$$

where $\exp(\eta_1) = \exp(\mu_1 + \sigma_1^2/2)$ and $\exp(\eta_2) = \exp(\mu_2 + \sigma_2^2/2)$ denote, respectively, the population mean of the refinery measurements and the mean of the BAAQMD measurements. For logged measurements taken by the refinery, we have: $n_1 = 31$, sample mean $\bar{y}_1 = 4.0743$ and $s_1 = 0.5021$; for logged measurements collected by the BAAQMD, $n_2 = 9$, $\bar{y}_2 = 2.963$ and $s_2 = 0.974$. To find the p-value for testing the above hypotheses using *StatCalc*, enter the sample sizes and the summary statistics, and click [p-values for] to get 0.112. Thus, we can not conclude that the true mean of the oil refinery measurements is greater than that of BAAQMD measurements. To get a 95% confidence intervals for the difference between two means using *StatCalc*, enter the sample sizes, the summary statistics and 0.95 for confidence level; click [2] to get $(-79.6, 57.3)$. To get one-sided limits, click [1]. The one-sided lower limit is -31.9 and the one-sided upper limit is 53.7.

22.7 Inferences for the Ratio of Two Means

Suppose that we have a sample of n_i observations from a lognormal population with parameters μ_i and σ_i, $i = 1, 2$. Let $\bar{y}_i$ and s_i denote, respectively, the mean and standard deviation of the logged measurements from the ith sample, $i = 1, 2$. For given $(n_1, \bar{y}_1, s_1, n_2, \bar{y}_2, s_2)$, *StatCalc* computes confidence intervals for the ratio $\exp(\eta_1)/\exp(\eta_2)$, where $\eta_i = \mu_i + \sigma_i^2/2$, $i = 1, 2$. *StatCalc* uses Algorithm 22.6.1 with $T_j = \exp(G_1)/\exp(G_2) = \exp(G_1 - G_2)$.

Example 22.7.1 Let us construct a 95% confidence interval for the ratio of the population means in Example 22.6.1. We have $n_1 = 31$ and $n_2 = 9$. For logged measurements, $\bar{y}_1 = 4.0743$, $s_1 = 0.5021$, $\bar{y}_2 = 2.963$ and $s_2 = 0.974$. To get a 95% confidence interval for the ratio of two means using *StatCalc*, select [StatCalc→ Continuous→Lognormal→CI for Mean1/Mean2], enter the sample sizes, the summary statistics and 0.95 for confidence level; click [2] to get (0.46, 4.16). Because this interval contains 1, we cannot conclude that the means are significantly different.

To get one-sided limits, click [1]. The one-sided lower limit is 0.67 and the one-sided upper limit is 3.75.

22.8 Applications

The lognormal distribution can be postulated in physical problems when the random variable X assumes only positive values and its histogram is remarkably skewed to the right. In particular, lognormal model is appropriate for a physical problem if the natural logarithmic transformation of the data satisfy normality assumption. Although lognormal and gamma distributions are interchangeable in many practical situations, a situation where they could produce different results is studied by Wiens (1999).

Practical examples where lognormal model is applicable vary from modeling raindrop sizes (Mantra and Gibbins 1999) to modeling the global position data (Kobayashi 1999). The latter article shows that the position data of selected vehicles measured by Global Positioning System (GPS) follow a lognormal distribution. Application of lognormal distribution in wind speed study is given in Garcia et. al. (1998) and Burlaga and Lazarus (2000). In exposure data analysis (data collected from employees who are exposed to workplace contaminants or chemicals) the applications of lognormal distributions are shown in Schulz and Griffin (1999), Borjanovic, et. al. (1999), Saltzman (1997), Nieuwenhuijsen (1997) and RoigNavarro, et. al. (1997). In particular, the one-sided tolerance limits of a lognormal distribution is useful in assessing the workplace exposure to toxic chemicals (Tuggle 1982). Wang and Wang (1998) showed that lognormal distributions fit very well to the fiber diameter data as well as the fiber strength data of merino wool. Lognormal distribution is also useful to describe the distribution of grain sizes (Jones et. al. 1999). Nabe et. al. (1998) analyzed data on inter-arrival time and the access frequency of world wide web traffic. They found that the document size and the request inter-arrival time follow lognormal distributions, and the access frequencies follow a Pareto distribution.

22.9 Properties and Results

The following results can be proved using the relation between the lognormal and normal distributions.

1. Let X_1 and X_2 be independent random variables with $X_i \sim \text{lognormal}(\mu_i, \sigma_i^2)$, $i = 1, 2$. Then

$$X_1 X_2 \sim \text{lognormal}(\mu_1 + \mu_2, \, \sigma_1^2 + \sigma_2^2)$$

   and

$$X_1/X_2 \sim \text{lognormal}(\mu_1 - \mu_2, \, \sigma_1^2 + \sigma_2^2).$$

2. Let $X_1, \ldots, X_n$ be independent lognormal random variables with parameters (μ, σ). Then

$$\text{Geometric Mean} = \left(\prod_{i=1}^{n} X_i\right)^{1/n} \sim \text{lognormal}\left(\mu, \, \frac{\sigma^2}{n}\right).$$

3. Let $X_1, \ldots, X_n$ be independent lognormal random variables with $X_i \sim \text{lognormal}(\mu_i, \, \sigma_i^2)$, $i = 1, \ldots, n$. For any positive numbers $c_1, \ldots, c_n$,

$$\prod_{i=1}^{n} c_i X_i \sim \text{lognormal}\left(\sum_{i=1}^{n} (\ln c_i + \mu_i), \, \sum_{i=1}^{n} \sigma_i^2\right).$$

For more results and properties, see Crow and Shimizu (1988).

22.10 Random Number Generation

Algorithm 22.10.1

For given μ and σ:
Generate z from N(0, 1)
Set $y = z * \sigma + \mu$
return $x = \exp(y)$

x is a pseudo random number from the lognormal(μ, σ^2) distribution.

22.11 Computation of Probabilities and Percentiles

Using the relation that

$$P(X \leq x) = P(\ln(X) \leq \ln(x)) = P\left(Z \leq \frac{\ln(x) - \mu}{\sigma}\right),$$

where Z is the standard normal random variable, the cumulative probabilities and the percentiles of a lognormal distribution can be easily computed. Specifically, if z_p denotes the pth quantile of the standard normal distribution, then $\exp(\mu + z_p\sigma)$ is the pth quantile of the lognormal(μ, σ^2) distribution.

Chapter 23

Pareto Distribution

23.1 Description

The probability density function of a Pareto distribution with parameters a and b is given by

$$f(x|a,b) = \frac{ba^b}{x^{b+1}},\ x \geq a > 0,\ b > 0. \tag{23.1.1}$$

The cumulative distribution function is given by

$$F(x|a,b) = P(X \leq x|a,b) = 1 - \left(\frac{a}{x}\right)^b,\ x \geq a. \tag{23.1.2}$$

For any given $0 < p < 1$, the inverse distribution function is

$$F^{-1}(p|a,b) = \frac{a}{(1-p)^{1/b}}. \tag{23.1.3}$$

Plots of the pdfs are given in Figure 23.1 for $b = 1, 2, 3$ and $a = 1$. All the plots show long right tail; this distribution may be postulated if the data exhibit a long right tail.

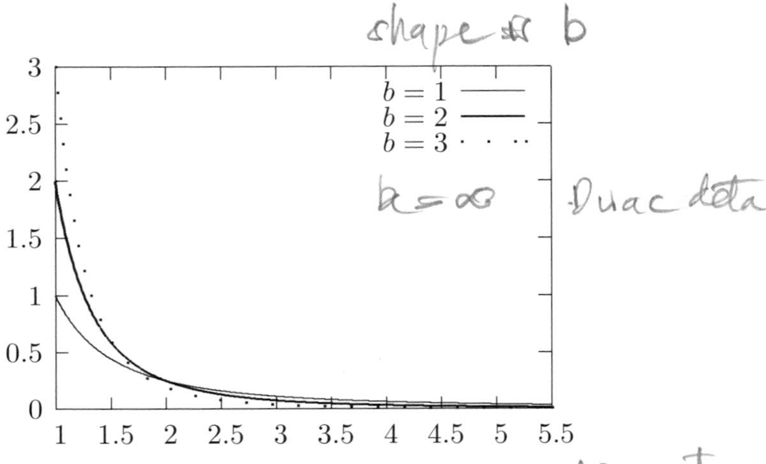

Figure 23.1 Pareto pdfs; $a = 1$

23.2 Moments

Mean:	$\frac{ab}{b-1}, \quad b > 1.$
Variance:	$\frac{ba^2}{(b-1)^2(b-2)}, \quad b > 2.$
Mode:	a
Median:	$a2^{1/b}$
Mean Deviation:	$\frac{2ab^{b-1}}{(b-1)^b}, \quad b > 1.$
Coefficient of Variation:	$\sqrt{\frac{1}{b(b-2)}}, \quad b > 2.$
Coefficient of Skewness:	$\frac{2(b+1)}{(b-3)}\sqrt{\frac{b-2}{b}}, \quad b > 3.$
Coefficient of Kurtosis:	$\frac{3(b-2)(3b^2+b+2)}{b(b-3)(b-4)}, \quad b > 4.$
Moments about the Origin:	$E(X^k) = \frac{ba^k}{(b-k)}, \quad b > k.$
Moment Generating Function:	does not exist.
Survival Function:	$(a/x)^b$
Hazard Function:	$b \ln(x/a)$

23.3 Computing Table Values

The dialog box [StatCalc→Continuous →Pareto] computes the cdf, percentiles, and moments of a Pareto(a, b) distribution.

To compute probabilities: Enter the values of the parameters a, b, and x; click [P(X <= x)].

Example 23.3.1 When $a = 2$, $b = 3$, and the value of $x = 3.4$, $P(X \leq 3.4) = 0.796458$ and $P(X > 3.4) = 0.203542$.

To compute percentiles: Enter the values of a, b, and the cumulative probability; click [x].

Example 23.3.2 When $a = 2$, $b = 3$, and the cumulative probability $= 0.15$, the 15th percentile is 2.11133. That is, $P(X \leq 2.11133) = 0.15$.

To compute other parameters: Enter the values of one of the parameters, cumulative probability and x. Click on the missing parameter.

Example 23.3.3 When $b = 4$, cumulative probability $= 0.15$, and $x = 2.4$, the value of a is 2.30444.

To compute moments: Enter the values a and b and click [M].

23.4 Inferences

Let $X_1, \ldots, X_n$ be a sample of independent observations from a Pareto(a, b) distribution with pdf in (23.1.1). The following inferences are based on the smallest order statistic $X_{(1)}$ and the geometric mean (GM). That is,

$$X_{(1)} = \min\{X_1, \ldots, X_n\} \quad \text{and} \quad GM = \left(\prod_{i=1}^{n} X_i\right)^{1/n}.$$

23.4.1 Point Estimation

Maximum Likelihood Estimators

$$\widehat{a} = X_{(1)}$$

and

$$\widehat{b} = \frac{1}{\ln(GM/\widehat{a})}.$$

Unbiased Estimators

$$\widehat{b}_u = \left(1 - \frac{1}{2n}\right)\widehat{b} \quad \text{and} \quad \widehat{a}_u = \left(1 - \frac{1}{(n-1)\widehat{b}}\right)\widehat{a},$$

where $\widehat{a}$ and $\widehat{b}$ are the MLEs given above.

23.4.2 Interval Estimation

A $1-\alpha$ confidence interval based on the fact that $2nb/\widehat{b} \sim \chi^2_{2(n-1)}$ is given by

$$\left(\frac{\widehat{b}}{2n}\chi^2_{2(n-1),\alpha/2}, \ \frac{\widehat{b}}{2n}\chi^2_{2(n-1),1-\alpha/2}\right).$$

If a is known, then a $1-\alpha$ confidence interval for b is given by

$$\left(\frac{\chi^2_{2n,\alpha/2}}{2n\ln(GM/a)}, \ \frac{\chi^2_{2n,1-\alpha/2}}{2n\ln(GM/a)}\right).$$

23.5 Applications

The Pareto distribution is often used to model the data on personal incomes and city population sizes. This distribution may be postulated if the histogram of the data from a physical problem has a long tail. Nabe et. al. (1998) studied the traffic data of world wide web (www). They found that the access frequencies of www follow a Pareto distribution. Atteia and Kozel (1997) showed that water particle sizes fit a Pareto distribution. The Pareto distribution is also used to describe the lifetimes of components. Aki and Hirano (1996) mentioned a situation where the lifetimes of components in a conservative-k-out-of-n-F system follow a Pareto distribution.

23.6 Properties and Results

1. Let $X_1, \ldots, X_n$ be independent Pareto(a, b) random variables. Then

   a.
   $$2b \ln\left(\frac{\prod_{i=1}^{n} X_i}{a^n}\right) \sim \chi^2_{2n}.$$

   b.
   $$2b \ln\left(\frac{\prod_{i=1}^{n} X_i}{(X_{(1)})^n}\right) \sim \chi^2_{2(n-1)},$$

   where $X_{(1)} = \min\{X_1, \ldots, X_n\}$.

23.7 Random Number Generation

For a given a and b:
Generate u from uniform(0, 1)
Set $x = a/(1-u) **(1/b)$

x is a pseudo random number from the Pareto(a, b) distribution.

23.8 Computation of Probabilities and Percentiles

Using the expressions for the cdf in (23.1.2) and inverse cdf in (23.1.3), the cumulative probabilities and the percentiles can be easily computed.

Chapter 24

Weibull Distribution

24.1 Description

Let Y be a standard exponential random variable with probability density function

$$f(y) = e^{-y}, \quad y > 0.$$

Define

$$X = bY^{1/c} + m, \quad b > 0, \ c > 0.$$

The distribution of X is known as the Weibull distribution with shape parameter c, scale parameter b, and the location parameter m. Its probability density is given by

$$f(x|b,c,m) = \frac{c}{b}\left(\frac{x-m}{b}\right)^{c-1} \exp\left\{-\left[\frac{x-m}{b}\right]^{c}\right\}, \ x > m, \ b > 0, \ c > 0. \tag{24.1.1}$$

The cumulative distribution function is given by

$$F(x|b,c,m) = 1 - \exp\left\{-\left[\frac{x-m}{b}\right]^{c}\right\}, \quad x > m, \ b > 0, \ c > 0. \tag{24.1.2}$$

For $0 < p < 1$, the inverse distribution function is

$$F^{-1}(p|b,c,m) = m + b(-\ln(1-p))^{\frac{1}{c}}. \tag{24.1.3}$$

Let us denote the three-parameter distribution by Weibull(b,c,m).

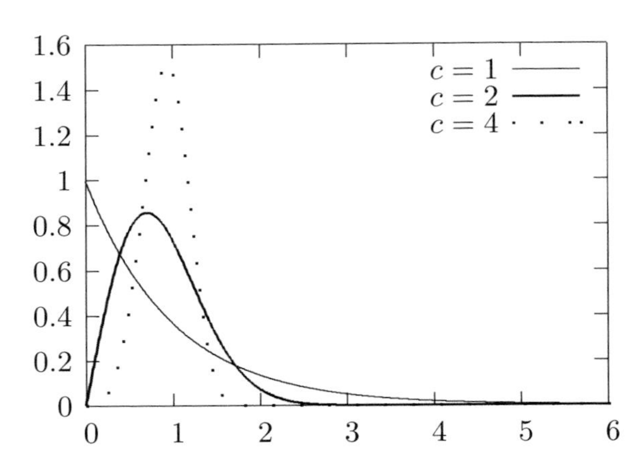

Figure 24.1 Weibull pdfs; $m = 0$ and $b = 1$

24.2 Moments

The following formulas are valid when $m = 0$.

Mean:	$b\Gamma(1 + 1/c)$
Variance:	$b^2\Gamma(1 + 2/c) - [\Gamma(1 + 1/c)]^2$
Mode:	$b\left(1 - \frac{1}{c}\right)^{1/c}, \ c \geq 1.$
Median:	$b[\ln(2)]^{1/c}$
Coefficient of Variation:	$\frac{\sqrt{\Gamma(1+2/c)-[\Gamma(1+1/c)]^2}}{\Gamma(1+1/c)}$
Coefficient of Skewness:	$\frac{\Gamma(1+3/c)-3\Gamma(1+1/c)\Gamma(1+2/c)+2[\Gamma(1+1/c)]^3}{\left[\Gamma(1+2/c)-\{\Gamma(1+1/c)\}^2\right]^{3/2}}$
Moments about the Origin:	$E(X^k) = b^k\Gamma(1 + k/c)$
Inverse Distribution Function (p):	$b\{-\ln(1 - p)\}^{1/c}$
Survival Function:	$P(X > x) = \exp\{-(x/b)^c\}$

Inverse Survival Function (p):	$b\{(1/c)\ln(-p)\}$
Hazard Rate:	cx^{c-1}/b^c
Hazard Function:	$(x/b)^c$

24.3 Computing Table Values

The dialog box [StatCalc→Continuous→Weibull] computes the cdf, percentiles, and moments of a Weibull(b, c, m) distribution.

To compute probabilities: Enter the values of m, c, b, and the cumulative probability; click [P(X <= x)].

Example 24.3.1 When $m = 0$, $c = 2.3$, $b = 2$, and $x = 3.4$, $P(X \leq 3.4) = 0.966247$ and $P(X > 3.4) = 0.033753$.

To compute percentiles: Enter the values of m, c, b, and the cumulative probability; click [x].

Example 24.3.2 When $m = 0$, $c = 2.3$, $b = 2$, and the cumulative probability $= 0.95$, the 95th percentile is 3.22259. That is, $P(X \leq 3.22259) = 0.95$.

To compute other parameters: Enter the values of any two of m, c, b, cumulative probability, and x. Click on the missing parameter.

Example 24.3.3 When $m = 1$, $c = 2.3$, $x = 3.4$, and the cumulative probability $= 0.9$, the value of b is 1.67004.

To compute moments: Enter the values of c and b and click [M]. The moments are computed assuming that $m = 0$.

24.4 Applications

The Weibull distribution is one of the important distributions in reliability theory. It is the distribution that received maximum attention in the past few decades. Numerous articles have been written demonstrating applications of the Weibull distributions in various sciences. It is widely used to analyze the cumulative loss of performance of a complex system in systems engineering. In general, it can be used to describe the data on waiting time until an event occurs.

In this manner, it is applied in risk analysis, actuarial science and engineering. Furthermore, the Weibull distribution has applications in medical, biological, and earth sciences. Arkai et. al. (1999) showed that the difference curve of two Weibull distribution functions almost identically fitted the isovolumically contracting left ventricular pressure-time curve. Fernandez et. al. (1999) modeled experimental data on toxin-producing Bacillus cereus strain isolated from foods by a Weibull distribution. The paper by Zobeck et. al. (1999) demonstrates that the Weibull distribution is an excellent choice to describe the particle size distribution of dust suspended from mineral sediment.

Although a Weibull distribution may be a good choice to describe the data on lifetimes or strength data, in some practical situations it fits worse than its competitors. For example, Korteoja et. al. (1998) reported that the Laplace distribution fits the strength data on paper samples better than the Weibull and extreme value distributions. Parsons and Lal (1991) showed that the extreme value distribution fits flexural strength data better than the Weibull distribution.

24.5 Point Estimation

Let $X_1, \ldots, X_n$ be a sample of observations from a Weibull distribution with known m. Let $Z_i = X_i - m$, where m is a known location parameter, and let $Y_i = \ln(Z_i)$. An asymptotically unbiased estimator of $\theta = (1/c)$ is given by

$$\widehat{\theta} = \frac{\sqrt{6}}{\pi} \sqrt{\frac{\sum\limits_{i=1}^{n} (Y_i - \bar{Y})^2}{n-1}}.$$

Further, the estimator is asymptotically distributed as normal with variance $= 1.1/(c^2 n)$ [Menon 1963]. When m is known, the MLE of c is the solution to the equation

$$\widehat{c} = \left[\sum_{i=1}^{n} Z_i^{\widehat{c}} Y_i \Big/ \sum_{i=1}^{n} Z_i^{\widehat{c}} - \bar{Y} \right]^{-1},$$

and the MLE of b is given by

$$\widehat{b} = \left(\frac{1}{n} \sum_{i=1}^{n} Z_i^{\widehat{c}} \right)^{1/\widehat{c}}.$$

24.6 Properties and Results

1. Let X be a Weibull(b, c, m) random variable. Then,

$$\left(\frac{X - m}{b}\right)^c \sim \text{exp}(1),$$

that is, the exponential distribution with mean 1.

2. It follows from (1) and the probability integral transform that

$$1 - \exp\left[-\left(\frac{X - m}{b}\right)^c\right] \sim \text{uniform}(0,\ 1),$$

and hence

$$X = m + b[-\ln(1 - U)]^{1/c} \sim \text{Weibull}(b, c, m),$$

where U denotes the uniform(0, 1) random variable.

24.7 Random Number Generation

For a given m, b, and c:
Generate u from uniform(0, 1)
return $x = m + b * (-\ln(1 - u)) * *(1/c)$

x is a pseudo random number from the Weibull(b, c, m) distribution.

24.8 Computation of Probabilities and Percentiles

The tail probabilities and percentiles can be easily computed because the analytical expressions for the cdf (24.1.2) and the inverse cdf (24.1.3) are very simple to use.

Chapter 25

Extreme Value Distribution

25.1 Description

The probability density function of the extreme value distribution with the location parameter a and the scale parameter b is given by

$$f(x|a,b) = \frac{1}{b}\exp[-(x-a)/b]\exp\{-\exp[-(x-a)/b]\},\ b > 0. \tag{25.1.1}$$

The cumulative distribution function is given by

$$F(x|a,b) = \exp\{-\exp[-(x-a)/b]\}, \quad -\infty < x < \infty,\ b > 0. \tag{25.1.2}$$

The inverse distribution function is given by

$$F^{-1}(p|a,b) = a - b\ln(-\ln(p)), \quad 0 < p < 1. \tag{25.1.3}$$

We refer to this distribution as extreme(a, b). The family of distributions of the form (25.1.2) is referred to as Type I family. Other families of extreme value distributions are:

Type II:

$$F(x|a,b) = \begin{cases} 0 \ \text{ for } \ x < a, \\ \exp\{-\left(\frac{x-a}{b}\right)^{-k}\} \ \text{ for } \ x \geq a,\ k > 0. \end{cases}$$

Type III:

$$F(x|a,b) = \begin{cases} \exp\{-\left(\frac{a-x}{b}\right)^{k}\} \ \text{ for } \ x \leq a,\ k > 0, \\ 1 \ \text{ for } \ x > a. \end{cases}$$

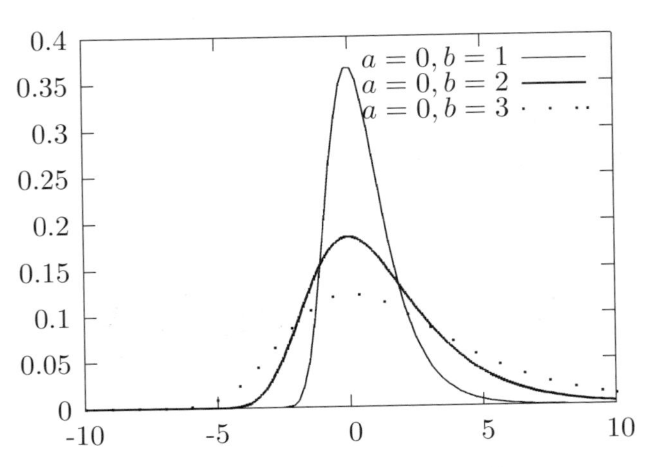

Figure 25.1 Extreme value pdfs

25.2 Moments

Mean:	$a + \gamma b$, where $\gamma = 0.5772\ 15664\ 9\ldots$.
Mode:	a
Median:	$a - b\ \ln(\ln 2)$
Variance:	$b^2\pi^2/6$
Coefficient of Skewness:	1.139547
Coefficient of Kurtosis:	5.4
Moment Generating Function:	$\exp(at)\,\Gamma(1 - bt), \quad t < 1/b.$
Characteristic Function:	$\exp(iat)\,\Gamma(1 - ibt)$
Inverse Distribution Function:	$a - b\ln(-\ln p)$
Inverse Survival Function:	$a - b\ \ln(-\ln\ (1 - p))$
Hazard Function:	$\frac{\exp[-(x-a)/b]}{b\{\exp[\exp(-(x-a)/b)]-1\}}$

25.3 Computing Table Values

The dialog box [StatCalc→Continuous→Extreme] computes probabilities, percentiles, and moments of an extreme value distribution.

To compute probabilities: Enter the values of the parameters a and b, and of x; click [P(X <= x)].

Example 25.3.1 When $a = 2$, $b = 3$ and $x = 2.3$, $P(X \leq 2.3) = 0.404608$ and $P(X > 2.3) = 0.595392$.

To compute percentiles: Enter the values of a, b and the cumulative probability; click [x].

Example 25.3.2 When $a = 1$, $b = 2$, and the cumulative probability $= 0.15$, the 15th percentile is -0.280674. That is, $P(X \leq -0.280674) = 0.15$.

Example 25.3.3 For any given three of the four values a, b, cumulative probability and x, *StatCalc* computes the missing one. For example, when $b = 2$, $x = 1$, and P(X <=x) = 0.15, the value of a is 2.28067.

To compute moments: Enter the values of a and b and click [M].

25.4 Maximum Likelihood Estimators

Let $X_1, ..., X_n$ be a sample from an extreme(a, b) distribution. Let $\bar{X}$ denote the sample mean, and $X_{(j)}$ denote the jth order statistic. The maximum likelihood estimators of b and a are given by

$$\widehat{b} = \bar{X} + \frac{1}{n}\sum_{j=1}^{n} X_{(j)}\left[\sum_{i=j}^{n}\frac{1}{i}\right]$$

and

$$\widehat{a} = -\widehat{b}\ln\left[\frac{1}{n}\sum_{i=1}^{n}\exp(-X_i/\widehat{b})\right].$$

25.5 Applications

Extreme value distributions are often used to describe the limiting distribution of the maximum or minimum of n observations selected from an exponential family of distributions such as normal, gamma, and exponential. They are also used to model the distributions of breaking strength of metals, capacitor breakdown voltage and gust velocities encountered by airplanes. Parsons and Lal (1991) studied thirteen sets of flexural strength data on different kinds of ice and found that between the three-parameter Weibull and the extreme value distributions, the latter fits the data better. Belzer and Kellog (1993) used the extreme value distribution to analyze the sources of uncertainty in forecasting peak power loads. Onoz and Bayazit (1995) showed that the extreme value distribution fits the flood flow data (collected from 1819 site-years from all over the world) best among seven distributions considered. Cannarozzo et al. (1995), Karim and Chowdhury (1995) and Sivapalan and Bloschl (1998) also used extreme value distributions to model the rainfall and flood flow data. Xu (1995) used the extreme value distribution to study the stochastic characteristics of wind pressures on the Texas Tech University Experimental Building.

Extreme value distributions are also used in stress-strength model. Harrington (1995) pointed out that if failure of a structural component is caused by the maximum of a sequence of applied loads, then the applied load distribution is an extreme value distribution. When strength of individual fibers is determined by the largest defect, an extreme value distribution describes the distribution of the size of the maximum defect of fibers. Lawson and Chen (1999) used an extreme value distribution to model the distribution of the longest possible microcracks in specimens of a fatigues aluminum-matrix silicon carbide whisker composite.

Kuchenhoff and Thamerus (1996) modeled extreme values of daily air pollution data by an extreme value distribution. Sharma et al. (1999) used an extreme value distribution for making predictions of the expected number of violations of the National Ambient Air Quality Standards as prescribed by the Central Pollution Control Board of India for hourly and eight-hourly average carbon monoxide concentration in urban road intersection region. Application of an extreme value distribution for setting the margin level in future markets is given in Longin (1999).

25.6 Properties and Results

1. If X is an exponential$(0, b)$, then $a - b\ln(X) \sim$ extreme$(a,\ b)$.
2. If X and Y are independently distributed as extreme$(a,\ b)$ random variable, then
$$X - Y \sim \text{logistic}(0,\ b).$$
3. If X is an extreme(0, 1) variable, then $b\exp(-X/c) \sim$ Weibull$(b, c, 0)$ and
$$\exp[-\exp(-X/b)] \sim \text{Pareto}(a,\ b).$$

25.7 Random Number Generation

For a given a and b:
Generate u from uniform(0, 1)
Set $x = a - b * \ln(-\ln(u))$

x is a pseudo random number from the extreme$(a,\ b)$ distribution.

25.8 Computation of Probabilities and Percentiles

The cumulative probabilities and percentiles of an extreme values distribution can be easily computed using (25.1.2) and (25.1.3).

Chapter 26

Cauchy Distribution

26.1 Description

The probability density function of a Cauchy distribution with the location parameter a and the scale parameter b is given by

$$f(x|a,b) = \frac{1}{\pi\, b[1 + ((x-a)/b)^2]}, \quad -\infty < a < \infty,\ b > 0.$$

The cumulative distribution function can be expressed as

$$F(x|a,b) = \frac{1}{2} + \frac{1}{\pi}\tan^{-1}\left(\frac{x-a}{b}\right), \quad b > 0. \tag{26.1.1}$$

We refer to this distribution as Cauchy(a, b). The standard forms of the probability density function and the cumulative distribution function can be obtained by replacing a with 0 and b with 1.

The inverse distribution function can be expressed as

$$F^{-1}(p|a,b) = a + b\tan(\pi(p-0.5)), \quad 0 < p < 1. \tag{26.1.2}$$

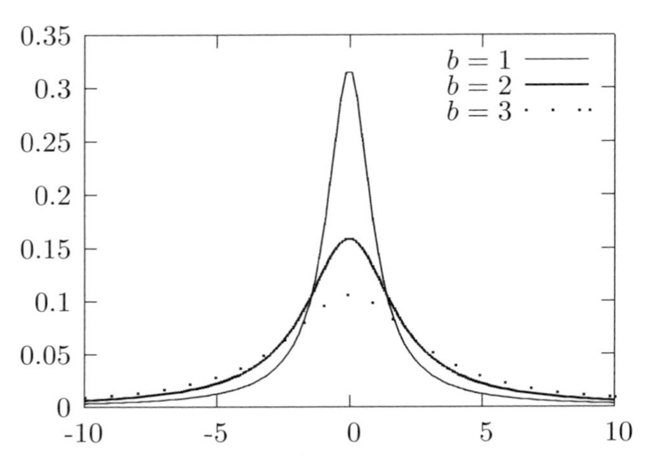

Figure 26.1 Cauchy pdfs; $a = 0$

26.2 Moments

Mean:	does not exist
Median:	a
Mode:	a
First Quartile:	$a - b$
Third Quartile:	$a + b$
Moments:	do not exist
Characteristic Function:	$\exp(ita - \|t\|b)$

26.3 Computing Table Values

The dialog box [StatCalc→Continuous→Cauchy] computes the cumulative probabilities and percentiles of a Cauchy distribution.

To compute probabilities: Enter the values of the parameters a and b, and of x; click [P(X <= x)].

Example 26.3.1 When $a = 1$, $b = 2$, and $x = 1.2$, $P(X \leq 1.2) = 0.531726$ and $P(X > 1.2) = 0.468274$.

To compute percentiles: Enter the values of a, b, and cumulative probability; click [x].

Example 26.3.2 When $a = 1$, $b = 2$ and the cumulative probability $= 0.95$, the 95th percentile is 13.6275. That is, $P(X \leq 13.6275) = 0.95$.

To compute parameters: Enter the value of one of the parameters, cumulative probability and x; click on the missing parameter.

Example 26.3.3 When $b = 3$, cumulative probability $= 0.5$, and $x = 1.25$, the value of a is 1.25.

26.4 Inference

Let $X_1, ..., X_n$ be a sample from a Cauchy(a, b) distribution. For $0.5 < p < 1$, let X_p and X_{1-p} denote the sample quantiles.

26.4.1 Estimation Based on Sample Quantiles

The point estimators of a and b based on the sample quantiles X_p and X_{1-p} and their variances are as follows.

$$\widehat{a} = \frac{X_p + X_{1-p}}{2}$$

with

$$\text{Var}(\widehat{a}) \simeq \frac{\widehat{b}^2}{n}\left[\frac{\pi^2}{2}(1-p)\right]\text{cosec}^4(\pi p), \qquad (26.4.1)$$

and

$$\widehat{b} = 0.5(x_p - x_{1-p})\tan[\pi(1-p)]$$

with

$$\text{Var}(\widehat{b}) \simeq \frac{\widehat{b}^2}{n}\left[2\pi^2(1-p)(2p-1)\right]\text{cosec}^2(2\pi p).$$

26.4.2 Maximum Likelihood Estimators

Maximum likelihood estimators of a and b are the solutions of the equations

$$\frac{1}{n}\sum_{i=1}^{n}\frac{2}{1+[(x_i-a)/b]^2}=1$$

and

$$\frac{1}{n}\sum_{i=1}^{n}\frac{2x_i}{1+[(x_i-a)/b]^2}=a.$$

26.5 Applications

The Cauchy distribution represents an extreme case and serves as counter examples for some well accepted results and concepts in statistics. For example, the central limit theorem does not hold for the limiting distribution of the mean of a random sample from a Cauchy distribution (see Section 26.6, Property 4). Because of this special nature, some authors consider the Cauchy distribution as a pathological case. However, it can be postulated as a model for describing data that arise as n realizations of the ratio of two normal random variables. Other applications given in the recent literature: Min et al. (1996) found that Cauchy distribution describes the distribution of velocity differences induced by different vortex elements. An application of the Cauchy distribution to study the polar and non-polar liquids in porous glasses is given in Stapf et al. (1996). Kagan (1992) pointed out that the Cauchy distribution describes the distribution of hypocenters on focal spheres of earthquakes. It is shown in the paper by Winterton et al. (1992) that the source of fluctuations in contact window dimensions is variation in contact resistivity, and the contact resistivity is distributed as a Cauchy random variable.

26.6 Properties and Results

1. If X and Y are independent standard normal random variables, then $X/Y \sim \text{Cauchy}(0,1)$.

2. If $X \sim \text{Cauchy}(0,1)$, then $2X/(1-X^2)$ also $\sim \text{Cauchy}(0, 1)$.

3. Student's t distribution specializes to the Cauchy$(0,1)$ distribution when df $= 1$.

4. If $X_1, \ldots, X_k$ are independent random variables with $X_j \sim \text{Cauchy}(a_j, b_j)$, $j = 1, \ldots, k$. Then

$$\sum_{j=1}^{k} c_j X_j \sim \text{Cauchy}\left(\sum_{j=1}^{k} c_j a_j,\ \sum_{j=1}^{k} |c_j|\, b_j\right).$$

5. It follows from (4) that the mean of a random sample of n independent observations from a Cauchy distribution follows the same distribution.

26.7 Random Number Generation

Generate u from uniform(0, 1)
Set $x = \tan[\pi * (u - 0.5)]$

x is a pseudo random number from the Cauchy(0, 1) distribution.

26.8 Computation of Probabilities and Percentiles

The cumulative probabilities and percentiles of a Cauchy distribution can be easily computed using (26.1.1) and (26.1.2), respectively.

Chapter 27

Inverse Gaussian Distribution

27.1 Description

The probability density function of X is given by

$$f(x|\mu,\sigma) = \left(\frac{\lambda}{2\pi x^3}\right)^{\frac{1}{2}} \exp\left(\frac{-\lambda(x-\mu)^2}{2\mu^2 x}\right), \quad x > 0,\ \lambda > 0,\ \mu > 0. \tag{27.1.1}$$

This distribution is usually denoted by IG(μ, λ). Using the standard normal cdf Φ, the cdf of an IG(μ, λ) can be expressed as

$$F(x|\mu,\lambda) = \Phi\left(\sqrt{\frac{\lambda}{x}}\left(\frac{x}{\mu}-1\right)\right) + e^{2\lambda/\mu}\Phi\left(-\sqrt{\frac{\lambda}{x}}\left(\frac{x}{\mu}+1\right)\right), \quad x > 0, \tag{27.1.2}$$

where $\Phi(x)$ is the standard normal distribution function.

Inverse Gaussian distributions offer a convenient modeling for positive right skewed data. The IG family is often used as alternative to the normal family because of the similarities between the inference methods for these two families.

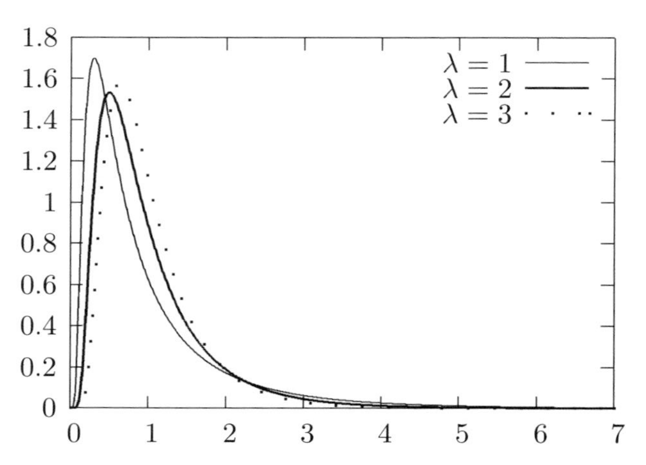

Figure 27.1 Inverse Gaussian pdfs; $\mu = 1$

27.2 Moments

Mean:	μ
Variance:	$\frac{\mu^3}{\lambda}$
Mode:	$\mu\left[\left(1+\frac{9\mu^2}{4\lambda^2}\right)^{1/2}-\frac{3\mu}{2\lambda}\right]$
Coefficient of Variation:	$\sqrt{\frac{\mu}{\lambda}}$
Coefficient of Skewness:	$3\sqrt{\frac{\mu}{\lambda}}$
Coefficient of Kurtosis:	$3+15\mu/\lambda$
Moments about the Origin:	$\mu^k\sum_{i=0}^{k-1}\frac{(k-1+i)!}{(k-1-i)!}\left(\frac{\mu}{2\lambda}\right)^i, \quad k\geq 2.$
Moment Generating Function	$\exp\left[\frac{\lambda}{\mu}\left(1-\left(1-\frac{2\mu^2 t}{\lambda}\right)^{1/2}\right)\right]$
Mean Deviation:	$4\sqrt{\frac{\lambda}{\mu}}\exp\left(2\sqrt{\frac{\lambda}{\mu}}\right)\Phi\left(-2\sqrt{\frac{\lambda}{\mu}}\right)\sqrt{\frac{\mu^3}{\lambda}}$, where Φ is the standard normal cdf.

27.3 Computing Table Values

The dialog box [StatCalc→Continuous→Inv Gau→Probabilities, Percentiles and Moments] computes the cumulative probabilities, percentiles, and moments of an IG(μ, λ) distribution. This dialog box also computes the necessary statistics for the following inferential methods.

1. Test and Confidence Interval for the Mean [Section 27.4].
2. Test and Confidence Interval for the Difference Between Two Means [Section 27.5.1].
3. Test and Confidence Interval for the Ratio of Two Means [Section 27.5.2].

To compute probabilities: Enter the values of the parameters μ and λ and the observed value x; click [P].

Example 27.3.1 When $\mu = 2$, $\lambda = 1$ and $x = 3$,

$$P(X \le 3) = 0.815981 \text{ and } P(X > 3) = 0.184019.$$

To compute percentiles: Enter the values of μ and λ, and the cumulative probability P(X <= x); click on [x].

Example 27.3.2 When $\mu = 1$, $\lambda = 2$, and the cumulative probability P(X <= x) = 0.95, the 95th percentile is 2.37739. That is, $P(X \le 2.37739) = 0.95$.

To compute moments: Enter the values of μ and λ; click [M].

27.4 One-Sample Inference

Let $X_1, \ldots, X_n$ be a sample from a IG(μ, λ) distribution. Define

$$\bar{X} = \frac{1}{n}\sum_{i=1}^{n} X_i \quad \text{and} \quad V = \frac{1}{n}\sum_{i=1}^{n}(1/X_i - 1/\bar{X}). \qquad (27.4.1)$$

The sample mean $\bar{X}$ is the MLE as well as unbiased estimate of μ and V^{-1} is the MLE of λ. $nV/(n-1)$ is the minimum variance unbiased estimator of $1/\lambda$.

The mean $\bar{X}$ and V are independent with $\bar{X} \sim \mathrm{IG}(\mu,\, n\lambda)$, and $\lambda n V \sim \chi^2_{n-1}$. Furthermore,

$$\left|\frac{\sqrt{(n-1)}(\bar{X}-\mu)}{\mu\sqrt{\bar{X}V}}\right| \sim |t_{n-1}|,$$

where t_m is a Student's t variable with df $= m$.

27.4.1 A Test for the Mean

Let $\bar{X}$ and V be as defined in (27.4.1). Define

$$S_1 = \sum_{i=1}^{n} (X_i + \mu_0)^2/X_i \quad \text{and} \quad S_2 = \sum_{i=1}^{n} (X_i - \mu_0)^2/X_i.$$

The p-value for testing

$$H_0: \mu \le \mu_0 \ \text{ vs. } \ H_a: \mu > \mu_0$$

is given by

$$F_{n-1}\left(-w_0\right) + \left(\frac{S_1}{S_2}\right)^{(n-2)/2} F_{n-1}\left(-\sqrt{4n + w_0^2\mu_0 S_1}\right), \tag{27.4.2}$$

where F_m denotes the cdf of Student's t variable with df $= m$, and w_0 is an observed value of

$$W = \frac{\sqrt{(n-1)}(\bar{X}-\mu_0)}{\mu_0\sqrt{\bar{X}V}}.$$

The p-value for testing

$$H_0: \mu = \mu_0 \ \text{ vs. } \ H_a: \mu \ne \mu_0,$$

is given by

$$P(|t_{n-1}| > |w_0|),$$

where t_m denotes the t variable with df $= m$.

27.4.2 Confidence Interval for the Mean

A $1-\alpha$ confidence interval for μ is given by

$$\left(\frac{\bar{X}}{1 + t_{n-1,1-\alpha/2}\sqrt{\frac{V\bar{X}}{n-1}}},\ \frac{\bar{X}}{\max\left\{0, 1 + t_{n-1,\alpha/2}\sqrt{\frac{V\bar{X}}{n-1}}\right\}}\right).$$

Example 27.4.1 Suppose that a sample of 18 observations from an IG(μ, λ) distribution yielded $\bar{X} = 2.5$ and $V = 0.65$. To find a 95% confidence interval for the mean μ, select [StatCalc→Continuous→Inv Gau→CI and Test for Mean], enter 0.95 for the confidence level, 18 for the sample size, 2.5 for the mean, and 0.65 for V, click [2-sided] to get (1.51304, 7.19009).

Suppose we want to test $H_0 : \mu = 1.4$ vs. $H_a : \mu \neq 1.4$. Enter 1.4 for [H0: M=M0], click [p-values for] to get 0.0210821.

27.5 Two-Sample Inference

The following two-sample inferential procedures are based on the generalized variable approach given in Krishnamoorthy and Tian (2005). This approach is valid only for two-sided hypothesis testing about $\mu_1 - \mu_2$, and constructing confidence intervals for $\mu_1 - \mu_2$ (not one-sided limits). More details can be found in the above mentioned paper.

Let $X_{i1}, ..., X_{in_i}$ be a sample from an IG(μ_i, λ_i) distribution, $i = 1, 2$. Let

$$\bar{X}_i = \frac{1}{n_i}\sum_{j=1}^{n_i} X_{ij} \quad \text{and} \quad V_i = \frac{1}{n_i}\sum_{j=1}^{n_i}(1/X_{ij} - 1/\bar{X}_i), i = 1, 2. \qquad (27.5.1)$$

The generalized variable of μ_i is given by

$$G_i = \frac{\bar{x}_i}{\max\left\{0, 1 + t_{n_i-1}\sqrt{\frac{\bar{x}_i v_i}{n_i-1}}\right\}}, \quad i = 1, 2, \qquad (27.5.2)$$

where $(\bar{x}_i, v_i)$ is an observed value of $(\bar{X}_i, V_i)$, $i = 1, 2$, and t_{n_1-1} and t_{n_2-1} are independent Student's t variables.

27.5.1 Inferences for the Difference between Two Means

Notice that for a given $(n_i, \bar{x}_i, v_i)$, the distribution of the generalized variable G_i in (27.5.2) does not depend on any unknown parameters. Therefore, Monte Carlo method can be used to estimate the p-value for testing about $\mu_1 - \mu_2$ or to find confidence intervals for $\mu_1 - \mu_2$. The procedure is given in the following algorithm.

Algorithm 27.5.1

For a given $(n_1, \bar{x}_1, v_1, n_2, \bar{x}_2, v_2)$:
For $j = 1, m$
Generate t_{n_1-1} and t_{n_2-1}
Compute G_1 and G_2 using (27.5.2)
Set $T_j = G_1 - G_2$
(end loop)

Suppose we are interested in testing

$$H_0 : \mu_1 = \mu_2 \text{ vs. } H_a : \mu_1 \neq \mu_2.$$

Then, the generalized p-value for the above hypotheses is given by

$$2 \min \{P(G_1 - G_2 < 0), P(G_1 - G_2 > 0)\}.$$

The null hypothesis will be rejected when the above p-value is less than a specified nominal level α. Notice that $P(G_1 - G_2 < 0)$ can be estimated by the proportion of the T_j's in Algorithm 27.5.1 that are less than zero; similarly $P(G_1 - G_2 > 0)$ can be estimated by the proportion of the T_j's that are greater than zero.

For a given $0 < \alpha < 1$, let T_α denotes the αth quantile of the T_j's in Algorithm 27.5.1. Then, $(T_{\alpha/2}, T_{1-\alpha/2})$ is a $1 - \alpha$ confidence interval for the mean difference.

StatCalc uses Algorithm 27.5.1 with $m = 1,000,000$ to compute the generalized p-value and generalized confidence interval. The results are almost exact (see Krishnamoorthy and Tian 2005).

Example 27.5.1 Suppose that a sample of 18 observations from an IG(μ_1, λ_1) distribution yielded $\bar{x}_1 = 2.5$ and $v_1 = 0.65$. Another sample of 11 observations from an IG(μ_2, λ_2) distribution yielded $\bar{x}_2 = 0.5$ and $v_2 = 1.15$. To find a 95% confidence interval for the mean difference $\mu_1 - \mu_2$, enter these statistics in the dialog box [StatCalc→Continuous→Inv Gau→CI and Test for Mean1-Mean2], 0.95 for the confidence level and click [2-sided] to get (0.85, 6.65).

The p-value for testing $H_0 : \mu_1 = \mu_2$ vs. $H_a : \mu_1 \neq \mu_2$ is given by 0.008.

?7.5.2 Inferences for the Ratio of Two Means

.et G_1 and G_2 be as defined in (27.5.2), and let $R = G_1/G_2$. The generalized)-value for testing

$$H_0 : \frac{\mu_1}{\mu_2} = 1 \quad \text{vs.} \quad H_a : \frac{\mu_1}{\mu_2} \neq 1$$

s given by

$$2 \min \{P(R < 1), P(R > 1)\}.$$

.et R_p denote the 100pth percentile of R. Then, $(R_{\alpha/2}, R_{1-\alpha/2})$ is a $1 - \alpha$:onfidence interval for the ratio of the IG means.

The generalized p-value and confidence limits for μ_1/μ_2 can be estimated ısing Monte Carlo method similar to the one given in Algorithm 27.5.1. The 'esults are very accurate for practical purposes (see Krishnamoorthy and Tian ?005).

Example 27.5.2 Suppose that a sample of 18 observations from an IG(μ_1, λ_1) listribution yielded $\bar{x}_1 = 2.5$ and $v_1 = 0.65$. Another sample of 11 observations 'rom an IG(μ_2, λ_2) distribution yielded $\bar{x}_2 = 0.5$ and $v_2 = 1.15$. To find a 95% :onfidence interval for the ratio μ_1/μ_2, enter these statistics and 0.95 for the conidence level in [StatCalc→ Continuous→Inv Gau→CI and Test for Mean1/Mean2], ınd click [2-sided] to get (2.05, 15.35).

The p-value for testing $H_0 : \mu_1 = \mu_2$ vs. $H_a : \mu_1 \neq \mu_2$ is given by 0.008.

?7.6 Random Number Generation

The following algorithm is due to Taraldsen and Lindqvist (2005).

Algorithm 27.6.1

'or a given μ and λ:
Generate $w \sim$ uniform$(0, 1)$ and $z \sim N(0, 1)$
set $v = z^2$
$d = \lambda/\mu$
$y = 1 - 0.5(\sqrt{v^2 + 4dv} - v)/d$
$x = y\mu$
if $(1 + y)w > 1$, set $x = 1/(y\mu)$

: is a random number from the IG(μ, λ) distribution.

27.7 Computational Methods for Probabilities and Percentiles

Since

$$P(X \le x|\mu,\lambda) = \Phi\left(\sqrt{\frac{\lambda}{x}}\left(\frac{x}{\mu}-1\right)\right) + e^{2\lambda/\mu}\Phi\left(-\sqrt{\frac{\lambda}{x}}\left(\frac{x}{\mu}+1\right)\right), \quad \text{for } x > 0,$$

the cumulative probabilities can be computed using the standard normal distribution function. *StatCalc* uses the above method for computing the cdf, and a root finding method to compute the percentiles.

Chapter 28

Rayleigh Distribution

28.1 Description

The Rayleigh distribution with the scale parameter b has the pdf

$$f(x|b) = \frac{x}{b^2}\exp\left(-\frac{1}{2}\frac{x^2}{b^2}\right), \quad x > 0,\ b > 0.$$

The cumulative distribution function is given by

$$F(x|b) = 1 - \exp\left(-\frac{1}{2}\frac{x^2}{b^2}\right), \quad x > 0,\ b > 0. \qquad (28.1.1)$$

Letting $F(x|b) = p$, and solving (28.1.1) for x, we get the inverse distribution function as

$$F^{-1}(p|b) = b\sqrt{-2\ln(1-p)}, \quad 0 < p < 1,\ b > 0. \qquad (28.1.2)$$

We observe from the plots of pdfs in Figure 28.1 that the Rayleigh distribution is always right skewed.

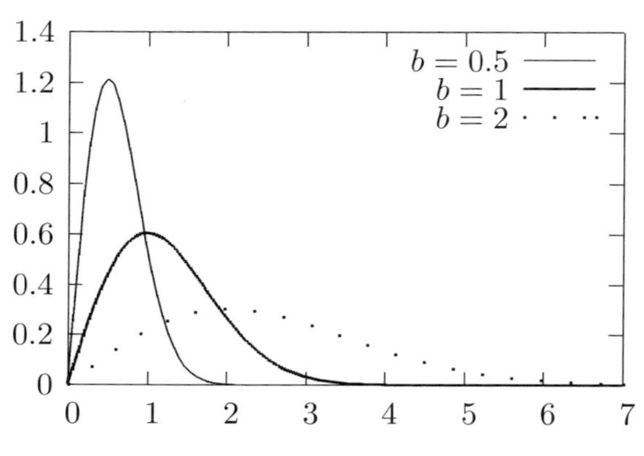

Figure 28.1 Rayleigh pdfs

28.2 Moments

Mean:	$b\sqrt{\frac{\pi}{2}}$
Variance:	$\left(2 - \frac{\pi}{2}\right) b^2$
Mode:	b
Median:	$b\sqrt{\ln(4)}$
Coefficient of Variation:	$\sqrt{(4/\pi - 1)}$
Coefficient of Skewness:	$\frac{2(\pi-3)\sqrt{\pi}}{(4-\pi)^{3/2}}$
Coefficient of Kurtosis:	$\frac{(32-3\pi^2)}{(4-\pi)^2}$
Moments about the Origin:	$2^{k/2}b^k\Gamma(k/2+1)$

28.3 Computing Table Values

The dialog box [StatCalc→Continuous→Rayleigh] computes the tail probabilities, percentiles and moments of a Rayleigh distribution.

To compute probabilities: Enter the values of the parameter b and x; click [P(X <= x)].

Example 28.3.1 When $b = 2$ and $x = 2.3$, $P(X \leq 2.3) = 0.733532$ and $P(X > 2.3) = 0.266468$.

To compute percentiles: Enter the values of b and the cumulative probability $P(X <= x)$; click on [x].

Example 28.3.2 When $b = 1.2$, and the cumulative probability $P(X <= x) = 0.95$, the 95th percentile is 2.07698. That is, $P(X \leq 2.07698) = 0.95$.

To compute the Value of b: Enter the values of x and the cumulative probability P(X <= x); click on [b].

Example 28.3.3 When $x = 3$, and the cumulative probability P(X <= x) = 0.9, the value of b is 1.97703. That is, $P(X \leq 3|b = 1.97703) = 0.9$.

To compute moments: Enter the value of b and click [M].

28.4 Maximum Likelihood Estimator

The MLE of b, based on a sample $X_1, ..., X_n$, is given by

$$\widehat{b} = \sqrt{\frac{1}{2n} \sum_{i=1}^{n} X_i^2}.$$

28.5 Relation to Other Distributions

1. Let X_1 and X_2 be independent N(0, b^2) random variables. Then, $Y = \sqrt{X_1^2 + X_2^2}$ follows a Rayleigh(b) distribution.

2. The Rayleigh(b) distribution is a special case of the Weibull distribution (see Chapter 24) with $b = \sqrt{2}b$, $c = 2$ and $m = 0$.

3. Let X be a Rayleigh(b) random variable. Then, $Y = X^2$ follows an exponential distribution with mean $2b^2$. That is, Y has the pdf

$$\frac{1}{2b^2} \exp\left(-\frac{y}{2b^2}\right), \quad y > 0.$$

28.6 Random Number Generation

Since the cdf has explicit from (see Section 28.1), random numbers can be generated using inverse transformation:

For a given b :
Generate u $\sim$ uniform(0,1)
Set $x = b * \sqrt{-2\ln(u)}$

x is a random number from the Rayleigh(b)

Chapter 29

Bivariate Normal Distribution

29.1 Description

Let (Z_1, Z_2) be a bivariate normal random vector with

$$E(Z_1) = 0, \ \mathrm{Var}(Z_1) = 1.0, \ E(Z_2) = 0, \ \mathrm{Var}(Z_2) = 1.0$$

and

$$\mathrm{Correlation}(Z_1, Z_2) = \rho.$$

The probability density function of (Z_1, Z_2) is given by

$$f(z_1, z_2|\rho) = \frac{1}{2\pi\sqrt{1-\rho^2}} \exp\left\{-\frac{1}{2(1-\rho^2)}\left(z_1^2 - 2\rho\, z_1 z_2 + z_2^2\right)\right\}, \quad (29.1.1)$$
$$-\infty < z_1 < \infty, \ -\infty < z_2 < \infty, \ -1 < \rho < 1.$$

Suppose that (X_1, X_2) is a bivariate normal random vector with

$$E(X_1) = \mu_1, \ \mathrm{Var}(X_1) = \sigma_{11}, \ E(X_2) = \mu_2, \ \mathrm{Var}(X_2) = \sigma_{22}$$

and the covariance, $\mathrm{Cov}(X_1, X_2) = \sigma_{12}$. Then

$$\left(\frac{X_1 - \mu_1}{\sqrt{\sigma_{11}}}, \frac{X_2 - \mu_2}{\sqrt{\sigma_{22}}}\right)$$

is distributed as (Z_1, Z_2) with correlation coefficient $\rho = \frac{\sigma_{12}}{\sqrt{\sigma_{11}\sigma_{22}}}$. That is,

$$P(X_1 \le a, X_2 \le b) = P\left(Z_1 \le \frac{a - \mu_1}{\sqrt{\sigma_{11}}}, Z_2 \le \frac{b - \mu_2}{\sqrt{\sigma_{22}}}\right).$$

The following relations are useful for computing probabilities over different regions.

a. $P(Z_1 \leq a,\ Z_2 > b) = \Phi(a) - P(Z_1 \leq a,\ Z_2 \leq b)$,

b. $P(Z_1 > a,\ Z_2 \leq b) = \Phi(b) - P(Z_1 \leq a,\ Z_2 \leq b)$,

c. $P(Z_1 > a,\ Z_2 > b) = 1 - \Phi(a) - \Phi(b) + P(Z_1 \leq a,\ Z_2 \leq b)$,

where Φ is the standard normal distribution function.

29.2 Computing Table Values

Let (X, Y) be a bivariate normal random vector with mean $= (0, 0)$, and the correlation coefficient ρ. For given x, y, and ρ, the dialog box [StatCalc→Continuous→Biv Normal→All Tail Probabilities] computes the following probabilities:

a. $P(X \leq x, Y > y)$

b. $P(X > x, Y > y)$

c. $P(X > x, Y \leq y)$

d. $P(X \leq x, Y \leq y)$, and

e. $P(|X| < x, |Y| < y)$.

Example 29.2.1 When $x = 1.1$, $y = 0.8$, and $\rho = 0.6$,

a. $P(X \leq 1.1, Y > 0.8) = 0.133878$

b. $P(X > 1.1, Y > 0.8) = 0.077977$

c. $P(X > 1.1, Y \leq 0.8) = 0.057689$

d. $P(X \leq 1.1, Y \leq 0.8) = 0.730456$, and

e. $P(|X| < 1.1, |Y| < 0.8) = 0.465559$.

If (X, Y) is a normal random vector with mean $= (\mu_1, \mu_2)$ and covariance matrix

$$\Sigma = \begin{pmatrix} \sigma_{11} & \sigma_{12} \\ \sigma_{21} & \sigma_{22} \end{pmatrix},$$

then to compute the probabilities at $(x,\, y)$, enter the standardized values $\frac{x-\mu_1}{\sqrt{\sigma_{11}}}$ for the x value, $\frac{y-\mu_2}{\sqrt{\sigma_{22}}}$ for the y value and $\frac{\sigma_{12}}{\sqrt{\sigma_{11}\sigma_{22}}}$ for the correlation coefficient, and click on [P].

29.3 An Example

Example 29.3.1 The Fuel Economy Guide published by the Department of Energy reports that for the 1998 compact cars the average city mileage is 22.8 with standard deviation 4.5, the average highway mileage is 31.1 with standard deviation is 5.5. In addition, the correlation coefficient between the city and highway mileage is 0.95.

a. Find the percentage of 1998 compact cars that give city mileage greater than 20 and highway mileage greater than 28.

b. What is the average city mileage of a car that gives highway mileage of 25?

Solution: Let (X_1, X_2) denote the (city, highway) mileage of a randomly selected compact car. Assume that (X_1, X_2) follows a bivariate normal distribution with the means, standard deviation and correlation coefficient given in the problem.

a.

$$\begin{aligned} P(X_1 > 20, X_2 > 28) &= P\left(Z_1 > \frac{20 - 22.8}{4.5},\ Z_2 > \frac{28 - 31.1}{5.5}\right) \\ &= P(Z_1 > -0.62,\ Z_2 > -0.56) \\ &= 0.679158. \end{aligned}$$

That is, about 68% of the 1998 compact cars give at least 20 city mileage and at least 28 highway mileage. To find the above probability, select the dialog box [StatCalc→Continuous→Biv Normal→All Tail Probabilities] from *StatCalc*, enter −0.62 for the [x value], −0.56 for the [y value], and 0.95 for the correlation coefficient; click [P].

b. From Section 29.5, Property 4, we have

$$\mu_1 + \sqrt{\frac{\sigma_{11}}{\sigma_{22}}}\rho(x_2 - \mu_2) = 22.8 + \frac{4.5}{5.5} \times 0.95 \times (25 - 31.1)$$
$$= 18.06 \text{ miles.}$$

For other applications and more examples, see "Tables of the Bivariate Normal Distribution Function and Related Functions," National Bureau of Standards, Applied Mathematics Series 50, 1959.

29.4 Inferences on Correlation Coefficients

Let $(X_{11}, X_{21}), \ldots, (X_{1n}, X_{2n})$ be a sample of independent observations from a bivariate normal population with

$$\text{covariance matrix } \Sigma = \begin{pmatrix} \sigma_{11} & \sigma_{12} \\ \sigma_{21} & \sigma_{22} \end{pmatrix}.$$

The population correlation coefficient is defined by

$$\rho = \frac{\sigma_{12}}{\sqrt{\sigma_{11}\sigma_{22}}}, \quad -1 \le \rho \le 1. \tag{29.4.1}$$

Define

$$\begin{pmatrix} \bar{X}_1 \\ \bar{X}_2 \end{pmatrix} = \begin{pmatrix} \frac{1}{n}\sum_{i=1}^{n} X_{1i} \\ \frac{1}{n}\sum_{i=1}^{n} X_{2i} \end{pmatrix} \quad \text{and} \quad S = \begin{pmatrix} s_{11} & s_{12} \\ s_{21} & s_{22} \end{pmatrix}, \tag{29.4.2}$$

where s_{ii} denotes the sample variance of X_i, $i = 1, 2$, and s_{12} denotes the sample covariance between X_1 and X_2 and is computed as

$$s_{12} = \frac{1}{n-1}\sum_{i=1}^{n}(X_{1i} - \bar{X}_1)(X_{2i} - \bar{X}_2).$$

The sample correlation coefficient is defined by

$$r = \frac{s_{12}}{\sqrt{s_{11}s_{22}}}, \quad -1 \le r \le 1.$$

The probability density function of r is given by

$$f(r|\rho) = \frac{2^{n-3}(1-\rho^2)^{(n-1)/2}(1-r^2)^{n/2-2}}{(n-3)!\pi}\sum_{i=0}^{\infty}\frac{(2r\rho)^i}{i!}\left\{\Gamma[(n+i-1)/2]\right\}^2. \tag{29.4.3}$$

Another form of the density function is given by Hotelling (1953):

$$\begin{aligned} &\frac{(n-2)}{\sqrt{2\pi}}\frac{\Gamma(n-1)}{\Gamma(n-1/2)}(1-\rho^2)^{(n-1)/2}(1-r^2)^{n/2-2}(1-r\rho)^{-n+3/2} \\ \times\quad & F\left(\frac{1}{2};\ \frac{1}{2};\ n-\frac{1}{2};\ \frac{1+r\rho}{2}\right), \end{aligned} \qquad (29.4.4)$$

where

$$F(a;b;c;x) = \sum_{j=0}^{\infty}\frac{\Gamma(a+j)}{\Gamma(a)}\frac{\Gamma(b+j)}{\Gamma(b)}\frac{\Gamma(c)}{\Gamma(c+j)}\frac{x^j}{j!}.$$

This series converges faster than the series in (29.4.3).

29.4.1 Point Estimation

The sample correlation coefficient r is a biased estimate of the population correlation coefficient ρ. Specifically,

$$E(r) = \rho - \frac{\rho(1-\rho^2)}{2(n-1)} + O\left(1/n^2\right).$$

The estimator

$$U(r) = r + \frac{r(1-r^2)}{(n-2)}, \qquad (29.4.5)$$

is an asymptotically unbiased estimator of ρ; the bias is of $O(1/n^2)$. [Olkin and Pratt 1958]

29.4.2 Hypothesis Testing

An Exact Test

Consider the hypotheses

$$H_0 : \rho \le \rho_0 \quad \text{vs.} \quad H_a : \rho > \rho_0. \qquad (29.4.6)$$

For a given n and an observed value r_0 of r, the test that rejects the null hypothesis whenever $P(r > r_0|n, \rho_0) < \alpha$ has exact size α. Furthermore, when

$$H_0 : \rho \ge \rho_0 \quad \text{vs.} \quad H_a : \rho < \rho_0, \qquad (29.4.7)$$

the null hypothesis will be rejected if $P(r < r_0|n, \rho_0) < \alpha$. The null hypothesis of

$$H_0 : \rho = \rho_0 \quad \text{vs.} \quad H_a : \rho \neq \rho_0, \qquad (29.4.8)$$

will be rejected whenever

$$P(r < r_0|n, \rho_0) < \alpha/2 \quad \text{or} \quad P(r > r_0|n, \rho_0) < \alpha/2.$$

The above tests are uniformly most powerful (UMP) among the scale and location invariant tests. (see Anderson 1984, p 114.) The above p-values can be computed by numerically integrating the pdf in (29.4.4).

A Generalized Variable Test

The generalized test given in Krishnamoorthy and Xia (2005) involves Monte Carlo simulation, and is equivalent to the exact test described above. The following algorithm can be used to compute the p-value and confidence interval for ρ.

Algorithm 29.4.1

For a given n and r:
  Set $r^* = r/\sqrt{1 - r^2}$
For $i = 1$ to m
  Generate $Z \sim N(0, 1)$, $U_1 \sim \chi^2_{n-1}$ and $U_2 \sim \chi^2_{n-2}$
  Set $G_i = \frac{r^*\sqrt{U_2} - Z}{\sqrt{\left(r^*\sqrt{U_2} - Z\right)^2 + U_1}}$
[end loop]

The generalized p-value for (29.4.6) is estimated by the proportion of G_i's that are less than ρ_0. The H_0 in (29.4.6) will be rejected if this generalized p-value is less than α. Similarly, the generalized p-value for (29.4.7) is estimated by the proportion of G_i's that are greater than ρ_0, and the generalized p-value for (29.4.8) is given by two times the minimum of one-sided p-values.

The dialog box [StatCalc→Continuous→Biv Normal→Test and CI for Correlation Coefficient] uses Algorithm 29.4.1 with $m = 500,000$ for computing the *generalized p-values* for hypothesis test about ρ. Krishnamoorthy and Xia (2005) showed that these generalized p-values are practically equivalent to the exact p-values described in the preceding subsection.

Example 29.4.1 Suppose that a sample of 20 observations from a bivariate normal population produced the correlation coefficient $r = 0.75$, and we like to test, $H_0 : \rho \leq 0.5$ vs. $H_a : \rho > 0.5$. To compute the p-value using *StatCalc*, enter 20 for n, 0.75 for [Sam Corrl r] and 0.5 for [rho_0]. Click [p-values for] to get 0.045. Since the p-value is less than 0.05, we can conclude that the population correlation coefficient is larger than 0.5 at the 5% level.

29.4.3 Interval Estimation

An Approximate Confidence Interval

An approximate confidence interval for ρ is based on the well-known Fisher's Z transformation of r. Let

$$Z = \frac{1}{2}\ln\left(\frac{1+r}{1-r}\right) \quad \text{and} \quad \mu_\rho = \frac{1}{2}\ln\left(\frac{1+\rho}{1-\rho}\right). \tag{29.4.9}$$

Then

$$Z \sim N(\mu_\rho, (n-3)^{-1}) \text{ asymptotically.} \tag{29.4.10}$$

The confidence interval for ρ is given by

$$\left(\tanh[Z - z_{\alpha/2}/\sqrt{n-3}],\ \tanh[Z + z_{\alpha/2}/\sqrt{n-3}]\right), \tag{29.4.11}$$

where $\tanh(x) = \frac{e^x - e^{-x}}{e^x + e^{-x}}$, and z_p is the upper pth quantile of the standard normal distribution.

An Exact Confidence Interval

Let r_0 be an observed value of r based on a sample of n bivariate normal observations. For a given confidence level $1-\alpha$, the upper limit ρ_U for ρ is the solution of the equation

$$P(r \leq r_0 | n, \rho_U) = \alpha/2, \tag{29.4.12}$$

and the lower limit $\rho_L$is the solution of the equation

$$P(r \geq r_0 | n, \rho_L) = \alpha/2. \tag{29.4.13}$$

See Anderson (1984, Section 4.2.2). One-sided limits can be obtained by replacing $\alpha/2$ by α in the above equations. Although (29.4.12) and (29.4.13) are difficult to solve for ρ_U and ρ_L, they can be used to assess the accuracy of the approximate confidence intervals (see Krishnamoorthy and Xia 2005).

The Generalized Confidence Interval for ρ

The generalized confidence interval due to Krishnamoorthy and Xia (2005) can be constructed using the percentiles of the G_i's given in Algorithm 29.4.1. Specifically, $(G_{\alpha/2}, G_{1-\alpha/2})$, where G_p, $0 < p < 1$, denotes the pth quantile of G_i's, is a $1 - \alpha$ confidence interval. Furthermore, these confidence intervals are exact because the endpoints satisfy the equations (29.4.13) and (29.4.12). For example, when $n = 20$, $r = 0.7$, the 95% generalized confidence interval for ρ (using *StatCalc*) is given by (0.365, 0.865). The probabilities in (29.4.13) and (29.4.12) are

$$P(r \geq 0.7|20, 0.365) = 0.025 \quad \text{and} \quad P(r \leq 0.7|20, 0.865) = 0.025.$$

In this sense, the generalized confidence limits are exact for samples as small as three (see Krishnamoorthy and Xia 2005).

The dialog box [StatCalc→Continuous→Biv Normal→Test and CI for Correlation Coefficient] uses Algorithm 29.4.1 with $m = 500,000$ for computing the *generalized confidence intervals* for ρ.

Example 29.4.2 Suppose that a sample of 20 observations from a bivariate normal population produced the correlation coefficient $r = 0.75$. To compute a 95% confidence interval for ρ using *StatCalc*, enter 20 for n, 0.75 for [Sam Corrl r] and 0.95 for [Conf Lev]. Click [1] to get one-sided confidence limits 0.509 and 0.872; click [2] to get confidence interval as (0.450, 0.889).

Example 29.4.3 The marketing manager of a company wants to determine whether there is a positive association between the number of TV ads per week and the amount of sales (in $1,000). He collected a sample of data from the records as shown in the following table.

Table 29.1 TV Sales Data

No. of ads, x_1	5	7	4	4	6	7	9	6	6
Sales, x_2	24	30	18	20	21	29	31	22	25

The sample correlation coefficient is given by

$$r = \frac{\sum_{i=1}^{n}(x_{1i} - \bar{x}_1)(x_{2i} - \bar{x}_2)}{\sqrt{\sum_{i=1}^{n}(x_{1i} - \bar{x}_1)^2 \sum_{i=1}^{n}(x_{2i} - \bar{x}_2)^2}} = 0.88.$$

An unbiased estimator using (29.4.5) is $0.88 + 0.88(1 - 0.88^2)/7 = 0.91$.

To compute a 95% confidence interval for the population correlation coefficient ρ, select [Continuous→Biv Normal→Test and CI for Correlation Coefficient] from *StatCalc*, enter 9 for n, 0.88 for sample correlation coefficient, 0.95 for the confidence level, click [2] to get (0.491, 0.969).

Suppose we want to test

$$H_0 : \rho \le 0.70 \quad \text{vs.} \quad H_a : \rho > 0.70.$$

To compute the p-value, enter 9 for n, 0.88 for the sample correlation and 0.70 for [rho_0]; click [p-values for] to get 0.122. Since this is not less than 0.05, at the 5% level, there is not sufficient evidence to indicate that the population correlation coefficient is greater than 0.70.

29.4.4 Inferences on the Difference between Two Correlation Coefficients

Let r_i denote the correlation coefficient based on a sample of n_i observations from a bivariate normal population with covariance matrix Σ_i, $i = 1, 2$. Let ρ_i denote the population correlation coefficient based on Σ_i, $i = 1, 2$.

An Asymptotic Approach

The asymptotic approach is based on Fisher's Z transformation given for the one-sample case, and is mentioned in Anderson (1984, p. 114). Let

$$Z_k = \frac{1}{2}\ln\left(\frac{1+r_k}{1-r_k}\right) \quad \text{and} \quad \mu_{\rho_k} = \frac{1}{2}\ln\left(\frac{1+\rho_k}{1-\rho_k}\right), \quad k = 1, 2. \qquad (29.4.14)$$

Then, it follows from the asymptotic result in (29.4.10) that,

$$\frac{(Z_1 - Z_2) - (\mu_{\rho_1} - \mu_{\rho_2})}{\sqrt{\frac{1}{n_1-3} + \frac{1}{n_2-3}}} \sim N(0, 1) \ \text{asymptotically.} \qquad (29.4.15)$$

Using the above asymptotic result one can easily develop test procedures for $\rho_1 - \rho_2$. Specifically, for an observed value (z_1, z_2) of (Z_1, Z_2), the p-value for testing

$$H_0 : \rho_1 \le \rho_2 \ \text{vs.} \ H_a : \rho_1 > \rho_2 \qquad (29.4.16)$$

is given by $1.0 - \Phi\left((z_1 - z_2)/\sqrt{1/(n_1-3) + 1/(n_2-3)}\right)$, where Φ is the standard normal distribution function. Notice that, using the distributional result in (29.4.15), one can easily obtain confidence interval for $\mu_{\rho_1} - \mu_{\rho_2}$ but not for $\rho_1 - \rho_2$.

Generalized Test and Confidence Limits for $\rho_1 - \rho_2$

The generalized p-values and confidence intervals can be obtained using the following algorithm.

Algorithm 29.4.2

For a given (r_1, n_1) and (r_2, n_2):
Set $r_1^* = r_1/\sqrt{1 - r_1^2}$ and $r_2^* = r_2/\sqrt{1 - r_2^2}$
For $i = 1$ to m
Generate $Z_{10} \sim N(0,1)$, $U_{11} \sim \chi^2_{n_1-1}$ and $U_{12} \sim \chi^2_{n_1-2}$
$Z_{20} \sim N(0,1)$, $U_{21} \sim \chi^2_{n_2-1}$ and $U_{22} \sim \chi^2_{n_2-2}$
Set $T_i = \frac{r_1^*\sqrt{U_{12}} - Z_{01}}{\sqrt{\left(r_1^*\sqrt{U_{12}} - Z_{01}\right)^2 + U_{11}}} - \frac{r_2^*\sqrt{U_{22}} - Z_{02}}{\sqrt{\left(r_2^*\sqrt{U_{22}} - Z_{02}\right)^2 + U_{21}}}$
(end loop)

Suppose we want to test

$$H_0 : \rho_1 - \rho_2 \le c \quad \text{vs.} \quad H_a : \rho_1 - \rho_2 > c, \tag{29.4.17}$$

where c is a specified number. The generalized p-value for (29.4.17) is estimated by the proportion of the T_i's that are less than c. The H_0 in (29.4.17) will be rejected if this generalized p-value is less than α. Similarly, the generalized p-value for a left-tail test is estimated by the proportion of T_i's that are greater than c, and the generalized p-value for a two-tail test is given by two times the minimum of the one-sided p-values.

Generalized confidence limits for $\rho_1 - \rho_2$ can be constructed using the percentiles of the T_i's in Algorithm 29.4.2. In particular, $(T_{\alpha/2},\ T_{1-\alpha/2})$, where T_p, $0 < p < 1$, is a $1 - \alpha$ confidence interval for $\rho_1 - \rho_2$.

The dialog box [StatCalc→Continuous→Biv Normal→Test and CI for rho1 - rho2] uses Algorithm 29.4.2 with $m = 500{,}000$ for computing the generalized p-values and generalized confidence intervals for $\rho_1 - \rho_2$.

Example 29.4.4 Suppose that a sample of 15 observations from a bivariate normal population produced the correlation coefficient $r_1 = 0.8$, and a sample from another bivariate normal population yielded $r_2 = 0.4$. It is desired to test

$$H_0 : \rho_1 - \rho_2 \le 0.1 \quad \text{vs.} \quad H_a : \rho_1 - \rho_2 > 0.1.$$

To compute the p-value using *StatCalc*, enter the sample sizes and correlation coefficients in the appropriate edit boxes, and 0.1 for [H0: rho1 - rho2]; click on [p-values for] to get 0.091. Because this p-value is not less than 0.05, the H_0 can not be rejected at the level of significance 0.05.

To get confidence intervals for $\rho_1 - \rho_2$, click on [1] to get one-sided limits, and click on [2] to get confidence interval. For this example, 95% one-sided limits for $\rho_1 - \rho_2$ are 0.031 and 0.778, and 95% confidence interval is $(-0.037, 0.858)$.

29.5 Some Properties

Suppose that (X_1, X_2) is a bivariate normal random vector with

$$E(X_1) = \mu_1, \ \text{Var}(X_1) = \sigma_{11}, \ E(X_2) = \mu_2, \ \text{Var}(X_2) = \sigma_{22}$$

and the covariance, $\text{Cov}(X_1, X_2) = \sigma_{12}$.

1. The marginal distribution of X_1 is normal with mean μ_1 and variance σ_{11}.
2. The marginal distribution of X_2 is normal with mean μ_2 and variance σ_{22}.
3. The distribution of $aX_1 + bX_2$ is normal with

$$\text{mean} = a\mu_1 + b\mu_2 \text{ and variance} = a^2\sigma_{11} + b^2\sigma_{22} + 2ab\sigma_{12}.$$

4. The conditional distribution of X_1 given X_2 is normal with

$$\begin{aligned} \text{mean} &= \mu_1 + \frac{\sigma_{12}}{\sigma_{22}}(x_2 - \mu_2) \\ &= \mu_1 + \rho\sqrt{\frac{\sigma_{11}}{\sigma_{22}}}(x_2 - \mu_2) \end{aligned}$$

and

$$\text{variance} = \sigma_{11} - \sigma_{12}^2/\sigma_{22}.$$

29.6 Random Number Generation

The following algorithm generates bivariate normal random vectors with mean (μ_1, μ_2) and variances σ_{11} and σ_{22}, and the correlation coefficient ρ.

Algorithm 29.5.1

Generate independent N(0, 1) variates u and v
Set
$x_1 = \mu_1 + \sqrt{\sigma_{11}} * u$
$x_2 = \sqrt{\sigma_{22}} * (\rho * u + v * \sqrt{1 - \rho * *2}) + \mu_2$

For a given n and ρ, correlation coefficients r can be generated using the following results. Let $A = (n-1)S$, where S is the sample covariance matrix defined in (29.4.2). Write

$$A = VV', \quad \text{where } V = \begin{pmatrix} v_{11} & 0 \\ v_{21} & v_{22} \end{pmatrix}, \quad v_{ii} > 0, \; i = 1, 2.$$

Similarly, let us write

$$\Sigma = \begin{pmatrix} \sigma_{11} & \sigma_{12} \\ \sigma_{21} & \sigma_{22} \end{pmatrix} = \theta\theta', \quad \text{where } \theta = \begin{pmatrix} \theta_{11} & 0 \\ \theta_{21} & \theta_{22} \end{pmatrix}, \quad \theta_{ii} > 0, \; i = 1, 2.$$

Then, V is distributed as θT, where T is a lower triangular matrix whose elements t_{ij} are independent with $t_{11}^2 \sim \chi_{n-1}^2$, $t_{22}^2 \sim \chi_{n-2}^2$ and $t_{21} \sim N(0,1)$. Furthermore, note that the population correlation coefficient can be expressed as

$$\rho = \frac{\theta_{21}}{\sqrt{\theta_{21}^2 + \theta_{22}^2}} = \frac{\theta_{21}/\theta_{22}}{\sqrt{\theta_{21}^2/\theta_{22}^2 + 1}}.$$

The above equation implies that

$$\frac{\theta_{21}}{\theta_{22}} = \rho \Big/ \sqrt{1 - \rho^2} = \rho^*, \text{ say.}$$

Similarly, the sample correlation coefficient can be expressed in terms of the elements v_{ij} of V as

$$r = \frac{v_{21}}{\sqrt{v_{21}^2 + v_{22}^2}} \sim \frac{\theta_{21}t_{11} + \theta_{22}t_{21}}{\sqrt{(\theta_{21}t_{11} + \theta_{22}t_{21})^2 + \theta_{22}^2 t_{22}^2}} = \frac{\rho^* t_{11} + t_{21}}{\sqrt{(\rho^* t_{11} + t_{21})^2 + t_{22}^2}}.$$

Using these results, we get the following algorithm for generating r.

Algorithm 29.5.2

For a given sample size n and ρ:

Set $\rho^* = \rho \Big/ \sqrt{1 - \rho^2}$
Generate $t_{11}^2 \sim \chi_{n-1}^2$, $t_{22}^2 \sim \chi_{n-2}^2$, and $t_{21} \sim N(0,1)$
Set $r = \frac{\rho^* t_{11} + t_{21}}{\sqrt{(\rho^* t_{11} + t_{21})^2 + t_{22}^2}}$

29.7 A Computational Algorithm for Probabilities

In the following $\Phi(x)$ denotes the standard normal distribution function and Prob $= P(Z_1 > x, Z_2 > y)$. Define

$$f(x,y) = \frac{1}{\sqrt{2\pi}} \int_0^x \Phi\left(\frac{ty}{x}\right) \exp(-t^2/2)dt.$$

If $\rho = 0$, return Prob $= (1 - \Phi(x)) * (1 - \Phi(y))$

If $x = 0$ and $y = 0$, return Prob $= 0.25 + \text{arc}\sin(\rho)/(2\pi)$

If $\rho = 1$ and $y \le x$, return Prob $= 1 - \Phi(x)$

If $\rho = 1$ and $y > x$, return Prob $= 1 - \Phi(y)$

If $\rho = -1$ and $x + y \ge 0$, return Prob $= 0$

If $\rho = -1$ and $x + y \le 0$, return Prob $= 1 - \Phi(x) - \Phi(y)$

$F = 0.25 - 0.5 * (\Phi(x) + \Phi(y) - 1) + \arcsin(\rho)/(2 * \pi)$

$\text{Prob} = f\left(x, \frac{y-\rho x}{\sqrt{1-\rho^2}}\right) + f\left(y, \frac{x-\rho y}{\sqrt{1-\rho^2}}\right) + F$

[Tables of the Bivariate Normal Distribution Function and Related Functions; National Bureau of Standards, Applied Mathematics Series 50, 1959]

Chapter 30

Distribution of Runs

30.1 Description

Consider a random arrangement of $m + n$ elements, m of them are one type, and n of them are of another type. A run is a sequence of symbols of the same type bounded by symbols of another type except for the first and last position. Let R denote the total number of runs in the sequence. The probability mass function of R is given by

$$P(R = r|m, n) = \frac{2\binom{m-1}{r/2-1}\binom{n-1}{r/2-1}}{\binom{m+n}{n}} \quad \text{for even } r,$$

and

$$P(R = r|m, n) = \frac{\binom{m-1}{(r-1)/2}\binom{n-1}{(r-3)/2} + \binom{m-1}{(r-3)/2}\binom{n-1}{(r-1)/2}}{\binom{m+n}{n}} \quad \text{for odd } r.$$

The distribution of runs is useful to test the hypothesis of randomness of an arrangement of elements. The hypotheses of interest are

$$H_0\text{: arrangement is random} \quad \text{vs.} \quad H_a\text{: arrangement is non random.}$$

Too many runs or too few runs provide evidence against the null hypothesis. Specifically, for a given m, n, the observed number of runs r of R, and the level of significance α, the H_0 will be rejected if the p-value

$$2\min\{P(R \le r), P(R \ge r)\} \le \alpha.$$

The plots of probability mass functions of runs in Figure 30.1 show that the run distribution is asymmetric when $m \neq n$ and symmetric when $m = n$.

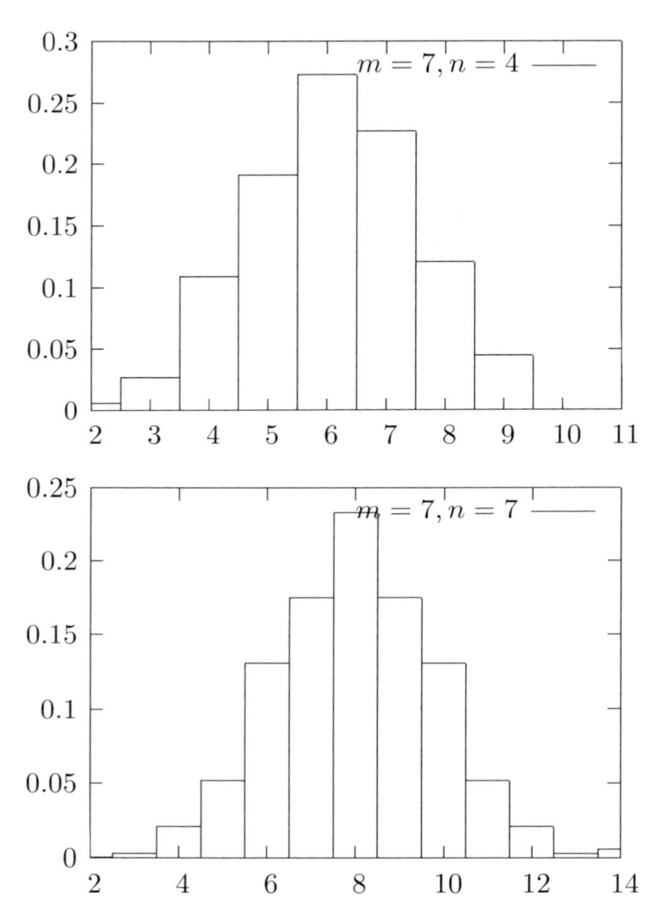

Figure 30.1 Probability Mass Functions of Runs

Moments

Mean: $1 + \frac{2mn}{m+n}$

Variance: $\frac{2mn(2mn-m-n)}{(m+n)^2(m+n-1)}$

30.2 Computing Table Values

For given m and n, the dialog box [StatCalc→Nonparametric→Distribution of Runs] evaluates the distribution function of R and critical points.

Example 30.2.1 Consider the following sequence:

$$\underline{a\;a\;a}\;\overline{b\;b}\;\underline{a}\;\overline{b\;b\;b}\;\underline{a\;a\;a\;a}\;\overline{b\;b}\;\underline{a\;a\;a\;a}\;\overline{b}$$

Here $m = 12$, $n = 8$, and the observed number of runs $r = 8$. To compute the probabilities using *StatCalc*, enter 12 for the number of [1st type symbols], 8 for the number of [2nd type symbols], and 8 for the [Observed runs r]; click on [P(R <= r)] to get

$$P(R \leq 8) = 0.159085 \quad \text{and} \quad P(R \geq 8) = 0.932603.$$

Since the probability of observing 8 or fewer runs is not less than 0.025, the null hypothesis that the arrangement is random will be retained at the level 0.05. To find the critical value, enter 0.025 for the tail probability, and click on [Left Crt] to get

$$P(R \leq 6) = 0.024609 \quad \text{and} \quad P(R \geq 16) = 0.00655.$$

This means that the null hypothesis of randomness will be rejected (at the 5% level) if the observed number of runs is 6 or less, or, 16 or more.

30.3 Examples

Example 30.3.1 Consider the simple linear regression model

$$Y_i = \alpha + \beta X_i + \epsilon_i, \quad i = 1, ..., N.$$

It is usually assumed that the error terms ϵ_i's are random. This assumption can be tested using the estimated errors

$$e_i = Y_i - \widehat{Y}_i, \quad i = 1, ..., N,$$

where Y_i and $\widehat{Y}_i$ denote, respectively, the observed value and the predicted value of the ith individual. Suppose that the estimated errors when $N = 18$ are:

-3.85	-0.11	6.63	-2.42	3.63	5.37	0.15	-2.76	-3.54
5.85	2.42	-6.37	-2.15	-0.37	3.24	3.48	-7.63	3.37

The arrangement of the signs of the errors is given by

$$- - + - + + + - - + + - - - + + - +$$

In this example, $m = 9$, $n = 9$, and the observed number of runs $r = 10$. We want to test the null hypothesis that the errors are randomly distributed at the level of significance 0.05. To compute the p-value, select the dialog box [StatCalc→Nonparametric→Distribution of Runs], enter 9 for the number of first type of symbols, 9 for the number of second type symbols, and 10 for the observed number of runs; click [P(R <= r)] to get

$$P(R \leq 10) = 0.600782 \quad \text{and} \quad P(R \geq 10) = 0.600782.$$

Since these probabilities are greater than $0.05/2 = 0.025$, the null hypothesis of randomness will be retained at the 5% level.

To get the critical values, enter 0.025 for the tail probability, and click [Left Crt] to get 5 and 15. That is, the left-tail critical value is 5 and the right-tail critical value is 15. Thus, the null hypothesis will be rejected at 0.05 level, if the total number of runs is 5 or less or 15 or more.

Example 30.3.2 Suppose that a sample of 20 students from a school is selected for some purpose. The genders of the students are recorded as they were selected:

M F M F F M F F M F M F F M F F M M F

We like to test if this really is a random sample. In this example, $m = 8$ (number of male students), $n = 12$ (number of female students), and the total number of runs $= 14$. Since the mean number of runs is given by

$$1 + \frac{2mn}{m+n} = 10.6,$$

it appears that there are too many runs, and so we want to compute the probability of observing 14 or more runs under the hypothesis of randomness. To compute the p-value using *StatCalc*, select the dialog box [StatCalc→Nonparametric→Distribution of Runs], enter 8 for the number of first type of symbols, 12 for the number of second type symbols, and 14 for the observed number of runs; click [P(R <= r)] to get $P(R \geq 14) = 0.0799$. Since the probability of observing 14 or more runs is not less than 0.025, we can accept the sample as a random sample at the 5% level.

Chapter 31

Sign Test and Confidence Interval for the Median

31.1 Hypothesis Test for the Median

Let $X_1, \ldots, X_n$ be a sample of independent observations from a continuous population. Let M denote the median of the population. We want to test

$$H_0 : M = M_0 \quad \text{vs.} \quad H_a : M \neq M_0.$$

Let K denote the number of plus signs of the differences $X_1 - M_0, \ldots, X_n - M_0$. That is, K is the number of observations greater than M_0. Then, under the null hypothesis, K follows a binomial distribution with number of trials n and success probability

$$P(X_i > M_0) = 0.5.$$

If K is too large, then we conclude that the true median $M > M_0$; if K is too small, then we conclude that $M < M_0$. Let k be an observed value of K. For a given level of significance α, the null hypothesis will be rejected in favor of the alternative hypothesis $M > M_0$ if $P(K \geq k|n, 0.5) \leq \alpha$, and in favor of the alternative hypothesis $M < M_0$ if $P(K \leq k|n, 0.5) \leq \alpha$. If the alternative hypothesis is $M \neq M_0$, then the null hypothesis will be rejected if

$$2 \min\{P(K \geq k|n, 0.5), P(K \leq k|n, 0.5)\} \leq \alpha.$$

31.2 Confidence Interval for the Median

Let $X_{(1)}, \ldots, X_{(n)}$ be the ordered statistics based on a sample $X_1, \ldots, X_n$. Let r be the largest integer such that

$$P(K \le r) = \sum_{i=0}^{r} \binom{n}{i} (0.5)^n \le \alpha/2 \tag{31.2.1}$$

and s is the smallest integer such that

$$P(K \ge s) = \sum_{i=s}^{n} \binom{n}{i} (0.5)^n \le \alpha/2. \tag{31.2.2}$$

Then, the interval $(X_{(r+1)}, X_{(s)})$ is a $1-\alpha$ confidence interval for the median M with coverage probability at least $1-\alpha$.

For a given sample size n and confidence level $1-\alpha$, the dialog box [StatCalc→Nonparametric→Sign Test and Confidence Interval for the Median] computes the integers r and s that satisfy (31.2.1) and (31.2.2) respectively.

Remark 31.2.1 If $X_i - M_0 = 0$ for some i, then simply discard those observations and reduce the sample size n accordingly. Zero differences can also be handled by assigning signs randomly (e.g. flip a coin; if the outcome is head, assign +, otherwise assign −).

31.3 Computing Table Values

The dialog box [StatCalc→Nonparametric→Signa Test and Confidence Interval for the Median] computes confidence intervals and p-values for testing the median.

Example 31.3.1 (*Confidence Interval*) To compute a 95% confidence interval for the median of a continuous population based on a sample of 40 observations, enter 40 for n, 0.95 for confidence level, click [CI] to get 14 and 27. That is, the required confidence interval is formed by the 14th and 27th order statistics from the sample.

Example 31.3.2 (*p-value*) Suppose that a sample of 40 observations yielded $k = 13$, the number of observations greater than the specified median. To obtain the p-value of the test when $H_a : M < M_0$, enter 40 for n, 13 for k and click [P(X <= k)] to get $P(K \le 13) = 0.0192387$. Since this p-value is less than 0.05, the null hypothesis that $H_0 : M = M_0$ will be rejected at the 5% level.

31.4 An Example

The nonparametric inferential procedures are applicable for any continuous distribution. However, in order to understand the efficiency of the procedures, we apply them to a normal distribution.

Example 31.4.1 A sample of 15 observations is generated from a normal population with mean 2 and standard deviation 1, and is given below.

1.01	3.00	1.12	1.68	0.82	4.01	2.85	2.49
1.58	2.30	2.84	2.32	3.01	1.77	2.10	

Recall that for a normal population, mean and median are the same. Let us test the hypotheses that

$$H_0 : M \leq 1.5 \quad \text{vs.} \quad H_a : M > 1.5.$$

Note that there are 12 data points are greater than 1.5. So, the p-value, $P(K \geq 12) = 0.01758$, which is less than 0.05. Therefore, the null hypothesis is rejected at the 5% level.

To find a 95% confidence interval for the median, enter 15 for the sample size n, and 0.95 for the confidence level. Click [CI] to get $(X_{(4)}, X_{(12)})$. That is, the 4th smallest observation and the 12th smallest (or the 4th largest) observation form the required confidence interval; for this example, the 95% confidence interval is (1.58, 2.84). Note that this interval indeed contains the actual median 2. Furthermore, if the hypotheses are

$$H_0 : M = 1.5 \quad \text{vs.} \quad H_a : M \neq 1.5,$$

then the null hypothesis will be rejected, because the 95% confidence interval does not contain 1.5.

Chapter 32

Wilcoxon Signed-Rank Test

32.1 Description

Wilcoxon signed-rank statistic is useful to test the median of a continuous symmetric distribution. Let M denote the median of the population. Consider the null hypothesis $H_0 : M = M_0$, where M_0 is a specified value of the median. Let $X_1, \ldots, X_n$ be a sample from the population and let

$$D_i = X_i - M_0, \quad i = 1, 2, \ldots, n.$$

Rank the absolute values $|D_1|, |D_2|, \ldots, |D_n|$. Let T^+ be the sum of the ranks of the positive $D_i's$. The distribution function of T^+ under H_0 is symmetric about its mean $n(n+1)/4$ (see Figure 32.1). The null hypothesis will be rejected if T^+ is too large or too small.

Let t be an observed value of T^+. For a given level α, the null hypothesis will be rejected in favor of the alternative

$$H_a : M > M_0 \quad \text{if} \quad P(T^+ \geq t) \leq \alpha,$$

and in favor of the alternative

$$H_a : M < M_0 \quad \text{if} \quad P(T^+ \leq t) \leq \alpha.$$

Furthermore, for a two-sided test, the null hypothesis will be rejected in favor of the alternative

$$H_a : M \neq M_0 \quad \text{if} \quad 2\min\{P(T^+ \leq t), P(T^+ \geq t)\} \leq \alpha.$$

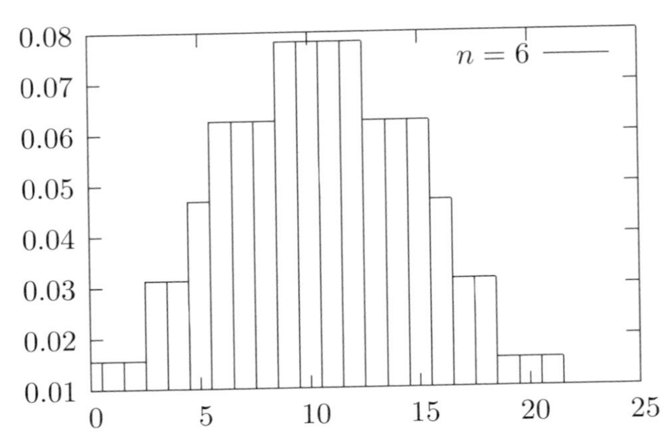

Figure 32.1 Probability Mass Function of WSR statistic

For a given n and an observed value t of T^+, the dialog box [StatCalc→ Nonparametric→Wilcoxon Signed-Rank Test] computes the above tail probabilities. *StatCalc* also computes the critical point for a given n and nominal level α.

For $\alpha < 0.5$, the left tail critical point k is the largest number such that

$$P(T^+ \leq k) \leq \alpha.$$

and the right tail critical point k is the smallest integer such that

$$P(T^+ \geq k) \leq \alpha.$$

32.2 Moments and an Approximation

Mean:	$\frac{n(n+1)}{4}$
Variance:	$\frac{n(n+1)(2n+1)}{24}$
	[Gibbons and Chakraborti 1992, p. 156]

An Approximation

The variable

$$Z = \frac{(T^+) \;- \text{mean}}{\sqrt{\text{var}}}$$

is approximately distributed as the standard normal random variable. Thi approximation is satisfactory for n greater than or equal to 15.

32.3 Computing Table Values

The dialog box [StatCalc→ Nonparametric→Wilcoxon Signed-Rank Test] computes the tail probabilities and critical points of the distribution of the signed-rank statistic T^+.

Example 32.3.1 Suppose that a sample of 20 observations yielded $T^+ = 130$. To compute tail probabilities using *StatCalc*, enter 20 for n, 130 for T^+, and click on [P(X <= k)] to get $P(T^+ \leq 130) = 0.825595$ and $P(T^+ \geq 130) = 0.184138$.

Example 32.3.2 Let $n = 20$. To get the upper 5% critical point, enter 20 for n, 0.95 for cumulative probability and click [Critical Pt] to get 150. That is, $P(T^+ \geq 150) = 0.048654$. To get the lower 5% critical point, enter 0.05 for cumulative probability and click [Critical Pt] to get 60. Notice that, because of the discreteness of the distribution, we get $P(T^+ \leq 60) = 0.048654$, not exactly 0.05.

32.4 An Example

Example 32.4.1 We will illustrate the Wilcoxon Signed-Rank test for the data given in Example 31.4.1. The data set consists of 15 observations generated from a normal population with mean 2 and standard deviation 1, and they are

1.01	3.00	1.12	1.68	0.82	4.01	2.85	2.49
1.58	2.30	2.84	2.32	3.01	1.77	2.10	

Note that for a normal population mean and median are the same. For illustration purpose, let us test the hypotheses that

$$H_0 : M \leq 1.5 \quad \text{vs.} \quad H_a : M > 1.5,$$

The differences $D_i = X_i - 1.5$ are

−0.49	1.5	−0.38	0.18	−0.68	2.51	1.30	
0.99	0.08	0.80	1.34	0.82	1.51	0.27	0.60

The ordered absolute values of D_i's with ranks are given below.

Rank	1	2	3	4	5	6	7	8
Ordered $\lvert D_i \rvert$	0.08	0.18	0.27	0.38*	0.49*	0.60	0.68*	0.80
Rank	9	10	11	12	13	14	15	
Ordered $\lvert D_i \rvert$	0.82	0.99	1.30	1.34	1.50	1.51	2.51	

The D_i's with negative sign are identified with *. Sum of the ranks of the positive differences can be computed as

$$\begin{aligned} T^+ &= \text{Total rank} - \text{Sum of the ranks of the negative } D_i'\text{s} \\ &= 15(15+1)/2 - 4 - 5 - 7 = 104. \end{aligned}$$

Using *StatCalc*, we can compute the p-value as

$$P(T^+ \geq 104) = 0.0051.$$

Since the p-value is less than any practical levels, we reject the null hypothesis there is sufficient evidence to indicate that the median is greater than 1.5. Note that the Wilcoxon Signed-Rank test provides stronger evidence against the nul hypothesis than the sign test (see Example 31.4.1).

To get the right-tail critical point using *StatCalc*, enter 15 for n, 0.95 for cumulative probability, and click [Critical Pt] to get

$$P(T^+ \geq 90) = 0.0473022.$$

Thus, any observed value of T^+ greater than or equal to 90 would lead to the rejection of H_0 at the 5% level of significance.

Chapter 33

Wilcoxon Rank-Sum Test

33.1 Description

The Wilcoxon rank-sum test is useful for comparing two continuous distributions. Let $X_1, \ldots, X_m$ be independent observations from a continuous distribution F_X and $Y_1, \ldots, Y_n$ be independent observations from a continuous distribution F_Y. Let W denote the sum of the ranks of the X observations in the combined ordered arrangement of the two samples. The range of W is given by

$$\frac{m(m+1)}{2} \leq W \leq \frac{m(2(m+n)-m+1)}{2}.$$

The W can be used as a test statistic for testing

$$H_0: F_X(x) = F_Y(x) \text{ for all } x \quad \text{vs.} \quad H_a: F_X(x) = F_Y(x-c) \ \text{ for some } c \neq 0,$$

The alternative hypothesis means that the distributions of X and Y are the same except for location. If W is too large or too small, then the null hypothesis will be rejected in favor of the alternative. If the hypotheses are

$$H_0: F_X(x) = F_Y(x) \text{ for all } x \quad \text{vs.} \quad H_a: F_X(x) = F_Y(x-c) \ \text{ for some } c > 0,$$

then the alternative hypothesis implies that the distribution of Y is shifted to the left from the distribution of X by c. That is, X values are more likely to be larger than Y values. In this case, the null hypothesis will be rejected for larger values of W. If the hypotheses are

$$H_0: F_X(x) = F_Y(x) \text{ for all } x \quad \text{vs.} \quad H_a: F_X(x) = F_Y(x+c) \quad \text{for some } c > 0,$$

then the alternative hypothesis implies that the distribution of Y is shifted to the right from the distribution of X by c. That is, X values are more likely to be smaller than Y values. In this case, the null hypothesis will be rejected for smaller values of W.

Let W_0 be an observed value of W and α be a specified level of significance. If $P(W \geq W_0) \leq \alpha$, then we conclude that $F_X(x) = F_Y(x - c),\ c > 0$; if $P(W \leq W_0) \leq \alpha$, then we conclude that $H_a : F_X(x) = F_Y(x + c),\ c > 0$. If $P(W \leq W_0) \leq \alpha/2$ or $P(W \geq W_0) \leq \alpha/2$, then we have evidence to conclude that $F_X(x) = F_Y(x - c),\ c \neq 0$.

33.2 Moments and an Approximation

Mean:	$\frac{m(m+n+1)}{2}$
Variance:	$\frac{mn(m+n+1)}{12}$
	[Gibbons and Chakraborti 1992, p. 241]

The variable

$$Z = \frac{W - \text{mean}}{\sqrt{\text{var}}} \sim N(0, 1) \text{ approximately.}$$

33.3 Mann-Whitney U Statistic

Wilcoxon rank-sum statistic and Mann-Whitney U statistic differ only by a constant, and, therefore, the test results based on these two statistics are the same. The Mann-Whitney U statistic is defined as follows: Let

$$D_{ij} = \begin{cases} 1 & \text{if } Y_j < X_i, \\ 0 & \text{otherwise,} \end{cases}$$

for $i = 1, \ldots, m$ and $j = 1, \ldots, n$. The statistic U is defined as

$$U = \sum_{i=1}^{m} \sum_{j=1}^{n} D_{ij},$$

or equivalently,

$$U = \sum_{i=1}^{n} (\text{number of } X's \text{ greater than the } Y_i).$$

t can be shown that

$$U = W - \frac{m(m+1)}{2},$$

nd hence

$$P(U \leq u) = P\left(W \leq u + \frac{m(m+1)}{2}\right).$$

33.4 Computing Table Values

The dialog box [StatCalc→Nonparametric →Wilcoxon Rank-Sum Test] computes probabilities and percentiles of the rank-sum statistic W.

To compute probabilities: Enter the values of m, n, and the observed value w; click on [P(W <= w)].

Example 33.4.1 When $m = 13$, $n = 12$ and the observed value w is 180, $P(W \leq 180) = 0.730877$ and $P(W \geq 180) = 0.287146$.

To compute the critical values: Enter the values of m, n, and the cumulative probability; click on [Critical Pt].

Example 33.4.2 When $m = 13$, $n = 12$, and the cumulative probability is 0.05, the left tail critical value is 138. Because of the discreteness of the distribution, the actual probability is 0.048821; that is $P(W \leq 138) = 0.048821$.

33.5 An Example

Example 33.5.1 It is desired to compare two types of treatments, A and B, based on recovery times of patients. Given below are recovery times of a sample of 9 patients who received treatment A, and recovery times of a sample of 9 patients who received treatment B.

Recovery times (in days)									
A:	17	19	20	24	13	18	21	22	25
B:	14	18	19	23	16	15	13	22	16

We want to see whether the data provide sufficient evidence to indicate that the recovery times of A are more likely longer than the recovery times of B. In notation,

$$H_0 : F_A(x) = F_B(x) \text{ for all } x, \quad \text{vs.} \quad H_a : F_A(x) = F_B(x - c), \qquad \text{for some } c > 0.$$

Note that the alternative hypothesis implies that the values of the random variable associated with A are more likely to be larger than those associated with B. The pooled sample data and their ranks (the average of the ranks is assigned to tied observations) are as follow.

Pooled sample and ranks

Treatment	B	A	B	B	B	B	A	A	B
data	13	13	14	15	16	16	17	18	18
ranks	1.5	1.5	3	4	5.5	5.5	7	8.5	8.5
Treatment	A	B	A	A	B	A	B	A	A
data	19	19	20	21	22	22	23	24	25
ranks	10.5	10.5	12	13	14.5	14.5	16	17	18

The sum of the ranks of A in the pooled sample is 102. The null hypothesis will be rejected if this sum is too large. Using [StatCalc→Nonparametric →Wilcoxon Rank-Sum Test], we get

$$P(W \geq 102) = 0.0807,$$

which is less than 0.1. Hence, the null hypothesis will be rejected at the 10% level.

Remark 33.5.1 *StatCalc* also computes critical values; for the above example, to compute right-tail 5% critical point, enter 9 for m, 9 for n and 0.95 for the cumulative probability; click [Critical Pt] to get 105. That is,

$$P(W \geq 105) = 0.04696.$$

Thus, for the above example, had the observed value of W been 105 or more, then we would have rejected the null hypothesis at the 5% level of significance.

Chapter 34

Nonparametric Tolerance Interval

34.1 Description

Let $X_1, \ldots, X_n$ be a sample from a continuous population. Let $X_{(k)}$ denote the kth order statistic; that is, the kth smallest of the X_i's. For a given p such that $0 < p < 1$, and g such that $0 < g < 1$, the dialog box [StatCalc→Nonparametric→Sample Size for NP Tolerance Interval] computes the value of n so that the interval

$$(X_{(1)}, X_{(n)})$$

would contain at least p proportion of the population with confidence level g. The interval $(X_{(1)}, X_{(n)})$ is called p content - g coverage tolerance interval or (p, g)–tolerance interval.

The required sample size n is the smallest integer such that

$$(n-1)p^n - np^{n-1} + 1 \geq g. \tag{34.1.1}$$

For a one-sided limit, the value of the n is the smallest integer such that

$$1 - p^n \geq g. \tag{34.1.2}$$

For a one-sided limit, the sample size is determined so that at least proportion p of the population data are greater than or equal to $X_{(1)}$; furthermore, at least proportion p of the population data are less than or equal to $X_{(n)}$.

Because of the discreteness of the sample size, the true coverage probability will be slightly more than the specified probability g. For example, when

$p = 0.90$ and $g = 0.90$, the required sample size for the two-sided tolerance limit is 38. Substituting 38 for n, 0.90 for p in (34.1.1) we get 0.9047, which is th actual coverage probability.

34.2 Computing Table Values

Select the dialog box [StatCalc→Nonparametric→Sample Size for NP Tolerance In terval] from *StatCalc*, enter the values of p and g; click on [Required Sample Siz n for].

Example 34.2.1 When $p = 0.90$ and $g = 0.95$, the value of n is 46; that i the interval $(X_{(1)}, X_{(46)})$ would contain at least 90% of the population wit confidence 95%. Furthermore, the sample size required for a one-sided toleranc limit is 29; that is 90% of the population data are less than or equal to $X_{(29}$ – the largest observation in a sample of 29 observations, with confidence 95% Similarly, 90% of the population data are greater than or equal to $X_{(1)}$ – th smallest observation in a sample of 29 observations, with confidence 95%.

34.3 An Example

Example 34.3.1 Suppose that one desires to find a tolerance interval so that i would contain 99% of household incomes in a large city with coverage probabil ity 0.95. How large should the sample be so that the interval (smallest orde statistic, largest order statistic) would contain at least 99% of the househol incomes with confidence 95%?

Solution: To compute the required sample size, enter 0.99 for p, 0.95 for g, an click [Required Sample Size] to get 473. That is, if we take a random sample o 473 households from the city and record their incomes, then at least 99% of th household incomes in the city fall between the lowest and the highest househol incomes in the sample with 95% confidence.

For one-sided limits, the required sample size is 299. That is, if we take a sample of 299 households from the city and record the incomes, then at least 99% of the household incomes are greater than the smallest income in the sample further, at least 99% of the household incomes are less than the highest incom in the sample.

Chapter 35

Tolerance Factors for a Multivariate Normal Population

35.1 Description

Let $X_1, \cdots, X_n$ be a sample of independent observations from an m-variate normal population with mean vector $\boldsymbol{\mu}$ and variance-covariance matrix $\boldsymbol{\Sigma}$. The sample mean vector and covariance matrix are computed as

$$\bar{X} = \frac{1}{n}\sum_{i=1}^{n} X_i \quad \text{and} \quad S = \frac{1}{n-1}\sum_{i=1}^{n}(X_i - \bar{X})(X_i - \bar{X})'.$$

For a given n, m, $0 < p < 1$, and $0 < g < 1$, the tolerance factor k is to be determined so that the region

$$\left\{X : (X - \bar{X})'S^{-1}(X - \bar{X}) \le k\right\}$$

would contain at least proportion p of the population with confidence g. Mathematically, k should be determined so that

$$P_{\bar{X},S}\left\{P_X[(X - \bar{X})'S^{-1}(X - \bar{X}) \le k|\bar{X}, S] \ge p\right\} = g,$$

where X follows the same m-variate normal distribution independently of the sample. We refer to this tolerance region as a $(p,\ g)$ tolerance region. At present no exact method of computing k is available. Krishnamoorthy and Mathew (1999) considered several approximation methods of computing k, and recommended the following one for practical use:

$$k \simeq \frac{d(n-1)\chi^2_{m,p}(\,m/n)}{\chi^2_{e,\,1-g}},$$

where $\chi^2_{m,p}(\delta)$ denotes the pth quantile of a noncentral chi-square distribution with df $= m$, and noncentrality parameter δ, $\chi^2_{e,\,1-g}$ denotes the $(1-g)$th quantile of a central chi-square distribution with df

$$e = \frac{4m(n-m-1)(n-m) - 12(m-1)(n-m-2)}{3(n-2) + m(n-m-1)},$$

and

$$d = \frac{e-2}{n-m-2}.$$

StatCalc computes this approximate tolerance factor k for a given n, m, p and g.

35.2 Computing Tolerance Factors

The dialog box [StatCalc→Miscellaneous→Tolerance Factors for a Mult.Variate Normal] computes tolerance factors for a multivariate normal distribution. To compute the tolerance factor, enter the sample size n, number of variables m, proportion p, and the coverage probability g; click on [Tol Factor].

Example 35.2.1 When $n = 35$, $m = 3$, $p = 0.90$ and $g = 0.95$, the 90% content – 95% coverage tolerance factor k is 9.83.

35.3 Examples

Example 35.3.1 A manufacturing company wants to setup a tolerance region for identifying skilled workers. A sample of workers is selected from a group of known skilled workers and the characteristics that are relevant to a skilled category are measured. These measurements can be used to construct, say, a (95%, 95%) tolerance region. This tolerance region may be used to classify a worker as skilled or non-skilled. For example, if the measurements on the characteristics of a new worker fall within the tolerance region, he or she may be classified as a skilled worker.

Example 35.3.2 Suppose that a lot of product is submitted for inspection. The buyer of the lot is interested in three characteristics (x_1, x_2, x_3) of the product (such as length, width, depth). A product is acceptable for the buyer's purpose if the measurements on $(x_1,\ x_2,\ x_3)$ of the product fall in a predetermined acceptable region. In order to save time and cost, the buyer may inspect a sample of product and construct a (95%, 95%), say, tolerance region for the lot. If the tolerance region falls within the acceptable region then the lot will be accepted.

Chapter 36

Distribution of the Sample Multiple Correlation Coefficient

36.1 Description

Let X be a m-variate normal random vector with unknown mean vector μ and unknown covariance matrix Σ. Partition X and Σ as

$$X = \begin{pmatrix} x_1 \\ X_2 \end{pmatrix} \quad \text{and} \quad \Sigma = \begin{pmatrix} \sigma_{11} & \Sigma_{12} \\ \Sigma_{21} & \Sigma_{22} \end{pmatrix},$$

where x_1 is the first component of X, and σ_{11} is the variance of x_1. The multiple correlation coefficient between the first component x_1 and the remaining $m-1$ components of X_2 is the maximum correlation between the first component and any linear combination of the other $m-1$ components, and is given by

$$\rho = \left(\frac{\Sigma_{12}\Sigma_{22}^{-1}\Sigma_{21}}{\sigma_{11}} \right)^{1/2}, \quad 0 \le \rho \le 1.$$

Let $X_1, \ldots, X_n$ be a sample of observations from an m-variate normal population. Define

$$\bar{X} = \frac{1}{n}\sum_{i=1}^{n} X_i, \quad \text{and} \quad S = \frac{1}{n-1}\sum_{i=1}^{n}(X_i - \bar{X})(X_i - \bar{X})'.$$

Partition S as

$$S = \begin{pmatrix} s_{11} & S_{12} \\ S_{21} & S_{22} \end{pmatrix},$$

so that s_{11} denotes the variance of the first component. The sample multiple correlation coefficient is defined by

$$R = \left(\frac{S_{12}S_{22}^{-1}S_{21}}{s_{11}}\right)^{1/2}, \quad 0 \le r \le 1.$$

The cumulative distribution function of R^2 is given by

$$P(R^2 \le x|\rho) = \sum_{k=0}^{\infty} b_k P\left(F_{m-1+2k,n-m} \le \frac{n-m}{m-1+2k}\left(\frac{x}{1-x}\right)\right), \tag{36.1.1}$$

where $F_{m-1+2k,n-m}$ denotes the F random variable, b_k is the negative binomial probability

$$b_k = \frac{((n-1)/2)_k}{k!}\left(\rho^2\right)^k (1-\rho^2)^{(n-1)/2}, \tag{36.1.2}$$

and $(a)_k = a(a-1)\cdots(a-k+1)$. The cumulative distribution is useful to test the hypotheses about ρ^2 or ρ and to find confidence intervals for ρ^2. [see Section 36.3]

36.2 Moments

Mean:

$$E(R^2) = \rho^2 + \frac{m-1}{n-1}(1-\rho^2) + \frac{2}{n+1}\rho^2(1-\rho^2) + O(1/n^2).$$

Variance:

$$\begin{aligned} \mathrm{Var}(R^2) &= \frac{4\rho^2(1-\rho^2)^2}{n} + O(1/n^2) \text{ when } \rho > 0, \\ &= \frac{2(n-m)(m-1)}{(n-1)^2(n+1)} \text{ when } \rho = 0. \end{aligned}$$

36.3 Inferences

36.3.1 Point Estimation

The square of the sample multiple correlation coefficient R^2 is a biased estimator of ρ^2. An asymptotically unbiased estimate of ρ^2 is given by

$$U(R^2) = R^2 - \frac{n-3}{n-m}(1-R^2) - \frac{2(n-3)}{(n-m)(n-m+2)}(1-R^2)^2.$$

The bias is of $O(1/n^2)$. [Olkin and Pratt 1958]

36.3.2 Interval Estimation

Let r^2 be an observed value of R^2 based on a sample of n observations. For a given confidence level $1-\alpha$, the upper limit U is the value of ρ^2 for which

$$P(R^2 \leq r^2 | n, U) = \alpha/2,$$

and the lower limit L is the value of ρ^2 for which

$$P(R^2 \geq r^2 | n, L) = \alpha/2.$$

The interval $(L,\ U)$ is an exact $1-\alpha$ confidence interval for ρ^2, and $(\sqrt{L},\ \sqrt{U})$ is an exact confidence interval for ρ. Kramer (1963) used this approach to construct table values for the confidence limits.

36.3.3 Hypothesis Testing

Consider the hypotheses

$$H_0 : \rho \leq \rho_0 \quad \text{vs.} \quad H_a : \rho > \rho_0.$$

For a given n and an observed value r^2 of R^2, the test that rejects the null hypothesis whenever

$$P(R^2 \geq r^2 | n, \rho_0^2) \leq \alpha$$

is a size α test. Furthermore, when

$$H_0 : \rho \geq \rho_0 \quad \text{vs.} \quad H_a : \rho < \rho_0,$$

the null hypothesis will be rejected if

$$P(R^2 \leq r^2 | n, \rho_0^2) \leq \alpha.$$

For testing

$$H_0 : \rho = \rho_0 \quad \text{vs.} \quad H_a : \rho \neq \rho_0,$$

will be rejected whenever

$$P(R^2 \leq r^2 | n, \rho_0^2) \leq \alpha/2 \quad \text{or} \quad P(R^2 \geq r^2 | n, \rho_0^2) \leq \alpha/2.$$

The above tests are uniformly most powerful among the invariant tests.

36.4 Some Results

1. Let $W = R^2/(1-R^2)$. Then,

$$P(W \le x|\rho) = \sum_{k=0}^{\infty} b_k P\left(F_{m-1+2k,n-m} \le \frac{n-m}{m-1+2k}x\right),$$

where b_k is the negative binomial probability

$$b_k = \frac{((n-1)/2)_k}{k!}\left(\rho^2\right)^k (1-\rho^2)^{(n-1)/2},$$

and $(a)_k = a(a-1)\cdots(a-k+1)$.

2. Let $\tau = \rho^2/(1-\rho^2)$, $a = \frac{(n-1)\tau(\tau+2)+m-1}{(n-1)\tau+m-1}$ and $b = \frac{((n-1)\tau+m-1)^2}{(n-1)\tau(\tau+2)+m-1}$.
Then,

$$P(R^2 \le x) \simeq P\left(Y \le \frac{x}{a(1-x)+x}\right),$$

where Y is a beta($b/2$, $(n-m)/2$) random variable. [Muirhead 1982, p. 176]

36.5 Random Number Generation

For a given sample size n, generate a Wishart random matrix A of order $m \times m$ with parameter matrix Σ, and set

$$R^2 = \frac{A_{12}A_{22}^{-1}A_{21}}{a_{11}}.$$

The algorithm of Smith and Hocking (1972) can be used to generate Wishart matrices.

36.6 A Computational Method for Probabilities

The following computational method is due to Benton and Krishnamoorthy (2003). The distribution function of R^2 can be written as

$$P(R^2 \le x) = \sum_{i=0}^{\infty} P(Y=i) I_x\left(\frac{m-1}{2}+i,\ \frac{v-m+1}{2}\right), \qquad (36.6.1)$$

where $v = n - 1$,

$$I_x(a,b) = \frac{\Gamma(a+b)}{\Gamma(a)\Gamma(b)} \int_0^x t^{a-1}(1-t)^{b-1}dt$$

is the incomplete beta function and

$$P(Y = i) = \frac{\Gamma(v/2+i)}{\Gamma(i+1)\Gamma(v/2)}\rho^{2i}(1-\rho^2)^{v/2}$$

is the negative binomial probability. Furthermore, $P(Y = i)$ attains its maximum around the integer part of

$$k = \frac{v\rho^2}{2(1-\rho^2)}.$$

To compute the cdf of R^2, first compute the kth term in (36.6.1) and then evaluate other terms using the following forward and backward recursions:

$$P(Y = i+1) = \frac{v/2+i}{i+1}\rho^2 P(Y = i), i = 0, 1, 2 \ldots,$$

$$P(Y = i-1) = \frac{i}{v/2+i-1}\rho^{-2} P(Y = i), i = 1, 2, \ldots,$$

$$I_x(a+1,b) = I_x(a,b) - \frac{\Gamma(a+b)}{\Gamma(a+1)\Gamma(b)}x^a(1-x)^b,$$

and

$$I_x(a-1,b) = I_x(a,b) + \frac{\Gamma(a+b-1)}{\Gamma(a+1)\Gamma(b)}x^{a-1}(1-x)^b.$$

The relation $\Gamma(a+1) = a\Gamma(a)$ can be used to evaluate the incomplete gamma function recursively. Forward and backward computations can be terminated if

$$1 - \sum_{j=k-i}^{k+i} P(Y = j)$$

is smaller than error tolerance or the number of iterations is greater than a specified number. Forward computations can be stopped if

$$\left(1 - \sum_{j=k-i}^{k+i} P(Y = j)\right) I_x\left(\frac{m-1}{2} + 2k + i + 1, \frac{v-m-1}{2}\right)$$

is less than or equal to error tolerance or the number of iterations is greater than a specified number.

36.7 Computing Table Values

To compute probabilities: Enter the values of the sample size n, number of variates m, squared population multiple correlation coefficient ρ^2, and the value of the squared sample multiple correlation coefficient r^2; click on $[P(X <= r^2)]$.

Example 36.7.1 When $n = 40$, $m = 4$, $\rho^2 = 0.8$ and $r^2 = 0.75$,

$$P(X \leq 0.75) = 0.151447 \text{ and } P(X > 0.75) = 0.84855.$$

To compute percentiles: Enter the values of the sample size n, number of variates m, squared population multiple correlation coefficient ρ^2, and the cumulative probability; click [Observed r^2].

Example 36.7.2 When $n = 40$, $m = 4$ and $\rho^2 = 0.8$, the 90th percentile is 0.874521.

To compute confidence intervals and p-values: Enter the values of n, m, r^2, and the confidence level; click [1-sided] to get one-sided limits; click [2-sided] to get confidence interval.

Example 36.7.3 Suppose that a sample of 40 observations from a four-variate normal population produced $r^2 = 0.91$. To find a 95% CI for the population squared multiple correlation coefficient, enter 40 for n, 4 for m, 0.91 for r^2, 0.95 for confidence level, and click [2-sided] to get (0.82102, 0.947821).

Suppose we want to test $H_0 : \rho^2 \leq 0.8$ vs. $H_a : \rho^2 > 0.8$. To find the p-value, enter 40 for n, 4 for m, 0.91 for r^2 and 0.8 for ρ^2; click [P(X <= r^2)] to get $P(X > 0.91) = 0.0101848$.

References

Abramowitz, M. and Stegun, I. A. (1965). *Handbook of Mathematical Functions.* Dover Publications, New York.

Aki, S. and Hirano, K. (1996). Lifetime distribution and estimation problems of consecutive–k–out–of–n: F systems. *Annals of the Institute of Statistical Mathematics*, 48, 185–199.

Anderson T. W. (1984). *An Introduction to Multivariate Statistical Analysis.* Wiley, New York.

Araki J., Matsubara H., Shimizu J., Mikane T., Mohri S., Mizuno J., Takaki M., Ohe T., Hirakawa M. and Suga, H. (1999). Weibull distribution function for cardiac contraction: integrative analysis. *American Journal of Physiology-Heart and Circulatory Physiology*, 277, H1940–H1945.

Atteia, O. and Kozel, R. (1997). Particle size distributions in waters from a karstic aquifer: from particles to colloids. *Journal of Hydrology*, 201, 102–119.

Bain, L. J. (1969). Moments of noncentral t and noncentral F distributions. *American Statistician*, 23, 33–34.

Bain, L. J. and Engelhardt, M. (1973). Interval estimation for the two-parameter double exponential distribution. *Technometrics*, 15, 875–887.

Balakrishnan, N. (ed.) (1991). *Handbook of the Logistic Distribution.* Marcel Dekker, New York.

Belzer, D. B. and Kellogg, M. A. (1993). Incorporating sources of uncertainty in forecasting peak power loads–a Monte Carlo analysis using the extreme value distribution. *IEEE Transactions on Power Systems*, 8, 730–737.

Benton, D. and Krishnamoorthy, K. (2003). Computing discrete mixtures of continuous distributions: noncentral chisquare, noncentral t and the distribution of the square of the sample multiple correlation coefficient. *Computational Statistics and Data Analysis*, 43, 249–267.

Benton, D., Krishnamoorthy, K. and Mathew, T. (2003). Inferences in multivariate–univariate calibration problems. *The Statistician* (JRSS-D), 52, 15–39.

Borjanovic, S. S., Djordjevic, S. V., Vukovic-Pal, M. D. (1999). A method for evaluating exposure to nitrous oxides by application of lognormal distribution. *Journal of Occupational Health*, 41, 27–32.

Bortkiewicz, L. von (1898). *Das Gesetz der Kleinen Zahlen.* Leipzig: Teubner.

Braselton, J., Rafter, J., Humphrey, P. and Abell, M. (1999). Randomly walking through Wall Street comparing lump-sum versus dollar-cost average investment strategies. *Mathematics and Computers in Simulation*, 49, 297–318.

Burlaga, L. F. and Lazarus A. J. (2000). Lognormal distributions and spectra of solar wind plasma fluctuations: Wind 1995–1998. *Journal of Geophysical Research-Space Physics*, 105, 2357–2364.

Burr, I. W. (1973). Some approximate relations between the terms of the hypergeometric, binomial and Poisson distributions. *Communications in Statistics–Theory and Methods*, 1, 297–301.

Burstein, H. (1975). Finite population correction for binomial confidence limits. *Journal of the American Statistical Association,* 70, 67–69.

Cacoullos, T. (1965). A relation between t and F distributions. *Journal of the American Statistical Association*, 60, 528–531.

Cannarozzo, M., Dasaro, F. and Ferro, V. (1995). Regional rainfall and flood frequency-analysis for Sicily using the 2–component extreme-value distribution. *Hydrological Sciences Journal-Journal des Sciences Hydrologiques*, 40, 19–42.

Chapman, D. G. (1952). On tests and estimates of the ratio of Poisson means. *Annals of the Institute of Statistical Mathematics*, 4, 45–49.

Chatfield, C., Ehrenberg, A. S. C. and Goodhardt, G. J. (1966). Progress on a simplified model of stationary purchasing behaviour. *Journal of the Royal Statistical Society, Series A*, 129, 317–367.

Chattamvelli, R. and Shanmugam, R. (1997). Computing the noncentral beta distribution function. *Applied Statistics*, 46, 146156.

Cheng, R. C. H. (1978). Generating beta variates with nonintegral shape parameters. *Communications* ACM, 21, 317-322.

Chhikara, R. S. and Folks, J. L. (1989). *The Inverse Gaussian Distribution.* Marcel Dekker, New York.

Chia, E. and Hutchinson, M. F. (1991). The beta distribution as a probability model for daily cloud duration. *Agricultural and Forest Meteorology*, 56, 195–208.

Clopper, C. J. and Pearsons E. S. (1934). The use of confidence or fiducial limits illustrated in the case of the binomial. *Biometrika*, 26, 404–413.

Craig, C. C. (1941). Note on the distribution of noncentral t with an application. *Annals of Mathematical Statistics*, 17, 193–194.

Crow, E. L. and Shimizu, K. (eds.) (1988). *Lognormal Distribution: Theory and Applications.* Marcel Dekker, New York.

Daniel, W. W. (1990). *Applied Nonparametric Statistics.* PWS-KENT Publishing Company, Boston.

Das, S. C. (1955). Fitting truncated type III curves to rainfall data. *Australian Journal of Physics*, 8, 298–304.

De Visor, C. L. M. and van den Berg, W. (1998). A method to calculate the size distribution of onions and its use in an onion growth model. *Sciatica Horticulturae*, 77, 129–143.

Fernandez A., Salmeron C., Fernandez P. S., Martinez A. (1999). Application of a frequency distribution model to describe the thermal inactivation of two strains of Bacillus cereus. *Trends in Food Science & Technology*, 10, 158–162.

Garcia, A., Torres, J. L., Prieto, E. and De Francisco, A. (1998). Fitting wind speed distributions: A case study. *Solar Energy*, 62, 139–144.

Gibbons, J. D. and Chakraborti, S. (1992). *Nonparametric Statistical Inference.* Marcel Dekker, New York.

Guenther, W. C. (1969). Shortest confidence intervals. *American Statistician*, 23, 22–25.

Guenther, W. C. (1971). Unbiased confidence intervals. *American Statistician*, 25, 18–20.

Guenther, W. C. (1978). Evaluation of probabilities for noncentral distributions and the difference of two t-variables with a desk calculator. *Journal of Statistical Computation and Simulation*, 6, 199–206.

Gumbel, E. J. and Mustafi, C. K. (1967). Some analytical properties of bivariate extremal distributions. *Journal of the American Statistical Association*, 62, 569–588.

Haff, L. R. (1979). An identity for the Wishart distribution with applications. *Journal of Multivariate Analysis*, 9, 531–544.

Harley, B. I. (1957). Relation between the distributions of noncentral t and a transformed correlation coefficient. *Biometrika*, 44, 219–224.

Hart, J.F., Cheney, E. W., Lawson, C. L., Maehly, H. J., Mesztenyi, H. J., Rice, J. R., Thacher, Jr., H. G., and Witzgall, C. (1968). *Computer Approximations.* John Wiley, New York.

Herrington, P. D. (1995). Stress-strength interference theory for a pin-loaded composite joint. *Composites Engineering*, 5, 975–982.

Hotelling, H (1953). New light on the correlation coefficient and its transforms. *Journal of the Royal Statistical Society B*, 15, 193–232.

Hwang, T. J. (1982). Improving on standard estimators in discrete exponential families with application to Poisson and negative binomial case. *The Annals of Statistics*, 10, 868–881.

Johnson, D. (1997). The triangular distribution as a proxy for the beta distribution in risk analysis. *Statistician*, 46, 387–398.

Jöhnk, M. D. (1964). Erzeugung von Betaverteilter und Gammaverteilter Zufallszahlen. *Metrika*, 8, 5–15.

Johnson, N. L. and Kotz, S. (1970). *Continuous univariate distributions - 2.* Houghton Mifflin Company, New York.

Johnson, N. L., Kotz, S. and Kemp, A. W. (1992). *Univariate Discrete Distributions.* John Wiley & Sons, New York.

Johnson, N. L., Kotz, S. and Balakrishnan, N. (1994). *Continuous Univariate Distributions.* Wiley, New York.

Jones, G. R. and Jackson M. and O'Grady, K. (1999). Determination of grain size distributions in thin films. *Journal of Magnetism and Magnetic Materials*, 193, 75–78.

Jonhk, M. D. (1964). Erzeugung von Betaverteilter und Gammaverteilter Zufallszahlen. *Metrika*, 8, 5–15.

Kachitvichyanukul, V. and Schmeiser, B. (1985). Computer generation of hypergeometric random variates. *Journal of Statistical Computation and Simulation*, 22, 127–145.

Kachitvichyanukul, V. and Schmeiser, B. (1988). Binomial random variate generation. *Communications of the ACM*, 31, 216–222.

Kagan, Y. Y. (1992). Correlations of earthquake focal mechanisms. *Geophysical Journal International*, 110, 305–320.

Kamat, A. R. (1965). Incomplete and absolute moments of some discrete distributions, classical and contagious discrete distributions, 45–64. Pergamon Press, Oxford.

Kappenman, R. F. (1977). Tolerance Intervals for the double-exponential distribution. *Journal of the American Statistical Association*, 72, 908–909.

Karim, M. A. and Chowdhury, J. U. (1995). A comparison of four distributions used in flood frequency-analysis in Bangladesh. *Hydrological Sciences Journal-Journal des Siences Hydrologiques*, 40, 55–66.

Kendall, M. G. (1943). *The Advance Theory of Statistics*, Vol. 1. Griffin, London.

Kendall, M. G. and Stuart, A. (1958). *The Advanced Theory of Statistics*, Vol. 1. Hafner Publishing Company, New York.

Kendall, M. G. and Stuart, A. (1973). *The Advanced Theory of Statistics*, Vol. 2. Hafner Publishing Company, New York.

Kennedy, Jr. W. J. and Gentle, J. E. (1980). *Statistical Computing.* Marcel Dekker, New York.

Kinderman, A. J. and Ramage, J. G. (1976). Computer generation of normal random variates. *Journal of the American Statistical Association*, 71, 893–896.

Kobayashi, T., Shinagawa, N. and Watanabe, Y. (1999). Vehicle mobility characterization based on measurements and its application to cellular communication systems. *IEICE transactions on communications*, E82B, 2055–2060.

Korteoja, M., Salminen, L. I., Niskanen, K. J., Alava, M. J. (1998). Strength distribution in paper. *Materials Science and Engineering a Structural Materials Properties Microstructure and Processing*, 248, 173–180.

Kramer, K. H. (1963). Tables for constructing confidence limits on the multiple correlation coefficient. *Journal of the American Statistical Association*, 58, 1082–1085.

Krishnamoorthy, K. and Mathew, T. (1999). Comparison of approximate methods for computing tolerance factors for a multivariate normal population. *Technometrics*, 41, 234–249.

Krishnamoorthy, K., Kulkarni, P. and Mathew, T. (2001). Hypothesis testing in calibration. *Journal of Statistical Planning and Inference*, 93, 211–223.

Krishnamoorthy, K. and Thomson, J. (2002). Hypothesis testing about proportions in two finite populations. *The American Statistician*, 56, 215–222.

Krishnamoorthy, K. and Mathew, T. (2003). Inferences on the means of lognormal distributions using generalized p-values and generalized confidence intervals. *Journal of Statistical Planning and Inference*, 115, 103 – 121.

Krishnamoorthy, K. and Mathew, T. (2004). One-sided tolerance limits in balanced and unbalanced one-way random models based on generalized confidence limits. *Technometrics*, (2004), 46, 44–52.

Krishnamoorthy, K. and Thomson, J. (2004). A more powerful test for comparing two Poisson means. *Journal of Statistical Planning and Inference*, 119, 23–35.

Krishnamoorthy, K. and Tian, L. (2004). Inferences on the difference between two inverse Gaussian means. Submitted for publication.

Krishnamoorthy, K. and Xia, Y. (2005). Inferences on correlation coefficients: one-sample, independent and correlated cases. Submitted for publication.

Kuchenhoff, H. and Thamerus, M. (1996). Extreme value analysis of Munich air pollution data. *Environmental and Ecological Statistics*, 3, 127–141.

Lawless, J. F. (1982). *Statistical Models and Methods for Lifetime Data.* Wiley, New York.

Lawson, L. R. and Chen, E. Y. (1999). Fatigue crack coalescence in discontinuously reinforced metal matrix composites: implications for reliability prediction. *Journal of Composites Technology & Research*, 21, 147–152.

Lenth, R. V. (1989). Cumulative distribution function of the noncentral t distribution. *Applied Statistics*, 38, 185–189.

Longing, F. M. (1999). Optimal margin level in futures markets: extreme price movements. *Journal of Futures Markets*, 19, 127–152.

Looney, S. W. and Gulledge Jr., T. R. (1985). Use of the correlation coefficient with normal probability plots. *American Statistician*, 39, 75–79.

Mantra, A. and Gibbins, C. J. (1999). Modeling of raindrop size distributions from multiwavelength rain attenuation measurements. *Radio Science*, 34, 657–666.

Majumder, K. L. and Bhattacharjee, G. P. (1973a). Algorithm AS 63. The incomplete beta integral. *Applied Statistics*, 22, 409–411.

Majumder, K. L. and Bhattacharjee, G. P. (1973b). Algorithm AS 64. Inverse of the incomplete beta function ratio. *Applied Statistics*, 22, 412–415.

Mathew, T. and Zha, W. (1996). Conservative confidence regions in multivariate calibration. *The Annals of Statistics*, 24, 707–725.

Menon, M. V. (1963). Estimation of the shape and scale parameters of the Weibull distribution. *Technometrics*, 5, 175–182.

Min, I. A., Mezic, I. and Leonard, A. (1996). Levy stable distributions for velocity and velocity difference in systems of vortex elements. *Physics of Fluids*, 8, 1169–1180.

Moser, B. K. and Stevens, G. R. (1992). Homogeneity of variance in the two-sample means test. *The American Statistician*, 46, 19-21.

Muirhead, R. J. (1982). *Aspects of Multivariate Statistical Theory.* John Wiley & Sons, New York.

Nabe, M., Murata, M. and Miyahara, H. (1998). Analysis and modeling of world wide web traffic for capacity dimensioning of internet access lines. *Performance Evaluation*, 34, 249–271.

Nicas, M. (1994). Modeling respirator penetration values with the beta distribution–an application to occupational tuberculosis transmission. *American Industrial Hygiene Association Journal*, 55, 515–524.

Nieuwenhuijsen, M. J. (1997). Exposure assessment in occupational epidemiology: measuring present exposures with an example of a study of occupational asthma. *International Archives of Occupational and Environmental Health*, 70, 295–308.

Odeh R. E., Owen, D. B., Birnbaum, Z. W. and Fisher, L. (1977). *Pocket Book of Statistical Tables.* Marcel Dekker, New York.

Olkin, I. and Pratt, J. (1958). Unbiased estimation of certain correlation coefficients. *The Annals of Mathematical Statistics*, 29, 201–211.

Onoz, B. and Bayazit, M. (1995). Best-fit distributions of largest available flood samples. *Journal of Hydrology*, 167, 195–208.

Oguamanam, D. C. D., Martin, H. R. and Huissoon, J. P. (1995). On the application of the beta distribution to gear damage analysis. *Applied Acoustics*, 45, 247–261.

Owen, D. B. (1968). A survey of properties and application of the noncentral t distribution. *Technometrics*, 10, 445–478.

Owen, D. B. (1964). Control of percentages in both tails of the normal distribution. *Technometrics*, 6, 377–387.

Parsons, B. L., and Lal, M. (1991). Distribution parameters for flexural strength of ice. *Cold Regions Science and Technology*, 19, 285–293.

Patel J. K., Kapadia, C. H. and Owen, D. B. (1976). *Handbook of Statistical Distributions*. Marcel Dekker, New York.

Patel, J. K. and Read, C. B. (1981). *Handbook of the Normal Distribution*. Marcel Dekker, New York.

Patil, G. P. (1962). Some methods of estimation for the logarithmic series distribution. *Biometrics*, 18, 68–75.

Patil G. P. and Bildikar, S. (1966). On minimum variance unbiased estimation for the logarithmic series distribution. *Sankhya, Ser. A*, 28, 239–250.

Patnaik, P. B. (1949). The noncentral chi-square and F-Distributions and their Applications. *Biometrika*, 36, 202–232.

Peizer, D. B. and Pratt, J. W. (1968). A normal approximation for binomial, F, beta, and other common related tail probabilities. *Journal of the American Statistical Association*, 63, 1416–1483.

Press, W. H., Teukolsky, S. A., Vetterling, W. T. and Flannery, B. P. (1997). *Numerical Recipes in C*. Cambridge University Press.

Prince, J. R., Mumma, C. G. and Kouvelis, A. (1988). Applications of the logistic distribution to film sensitometry in nuclear-medicine. *Journal of Nuclear Medicine*, 29, 273–273.

Prochaska, B. J. (1973). A note on the relationship between the geometric and exponential distributions. *The American Statistician*, 27, 27.

Puri, P. S. (1973). On a property of exponential and geometric distributions and its relevance to multivariate failure rate. *Sankhya A*, 35, 61–78.

Rajkai, K., Kabos, S., VanGenuchten, M. T. and Jansson, P. E. (1996). Estimation of water-retention characteristics from the bulk density and particle-size distribution of Swedish soils. *Soil Science*, 161, 832–845.

Roig-Navarro, A. F., Lopez, F. J., Serrano, R. and Hernandez, F. (1997). An assessment of heavy metals and boron contamination in workplace atmospheres from ceramic factories. *Science of the Total Environment*, 201, 225–234.

Rutherford, E. and Geiger, H. (1910). The probability variations in the distribution of α particles. *Philosophical Magazine*, 20, 698–704.

Schmeiser, B. W. and Lal, L. (1980). Squeeze methods for generating gamma variates. *Journal of the American Statistical Association*, 75, 679–682.

Schmeiser, B. W. and Shalaby, M. A. (1980). Acceptance/Rejection methods for beta variate generation. *Journal of the American Statistical Association*, 75, 673–678.

Sahli, A., Trecourt, P. and Robin, S. (1997). One sided acceptance sampling by variables: The case of the Laplace distribution. *Communications in Statistics-Theory and Methods*, 26, 2817–2834 1997.

Saltzman, B. E. (1997). Health risk assessment of fluctuating concentrations using lognormal models. *Journal of the Air & Waste Management Association*, 47, 1152– 1160.

Scerri, E. and Farrugia, R. (1996). Wind data evaluation in the Maltese Islands. *Renewable Energy*, 7, 109–114.

SchwarzenbergCzerny, A. (1997). The correct probability distribution for the phase dispersion minimization periodogram. *Astrophysical Journal*, 489, 941–945.

Schulz, T. W. and Griffin, S. (1999). Estimating risk assessment exposure point concentrations when the data are not normal or lognormal. *Risk Analysis*, 19, 577–584.

Sharma, P., Khare, M. and Chakrabarti, S. P. (1999) Application of extreme value theory for predicting violations of air quality standards for an urban road intersection. *Transportation Research Part D–Transport and Environment*, 4, 201–216.

Shivanagaraju, C., Mahanty, B., Vizayakumar, K. and Mohapatra, P. K. J. (1998). Beta-distributed age in manpower planning models. *Applied Mathematical Modeling*, 22, 23–37.

Sivapalan ,M. and Bloschl, G. (1998). Transformation of point rainfall to areal rainfall: Intensity-duration frequency curves. *Journal of Hydrology*, 204, 150–167.

Smith, W. B. and Hocking, R. R. (1972). Wishart variates generator, Algorithm AS 53. *Applied Statistics*, 21, 341–345.

Stapf, S., Kimmich, R., Seitter, R. O., Maklakov, A. I. and Skid, V. D. (1996). Proton and deuteron field-cycling NMR relaxometry of liquids confined in porous glasses. *Colloids and Surfaces: A Physicochemical and Engineering Aspects*, 115, 107–114.

Stein, C. (1981). Estimation of the mean of a multivariate normal distribution. *The Annals of Statistics*, 9, 1135–1151.

Storer, B. E., and Kim, C. (1990). Exact properties of some exact test statistics comparing two binomial proportions. *Journal of the American Statistical Association*, 85, 146—155.

Stephenson, D. B., Kumar, K. R., Doblas-Reyes, F. J., Royer, J. F., Chauvin, E. and Pezzulli, S. (1999). Extreme daily rainfall events and their impact on ensemble forecasts of the Indian monsoon. *Monthly Weather Review*, 127, 1954–1966.

Sulaiman, M. Y., Oo, W. H., Abd Wahab, M., Zakaria, A. (1999). Application of beta distribution model to Malaysian sunshine data. *Renewable Energy*, 18, 573–579.

Taraldsen, G. and Lindqvist, B. (2005). The multiple roots simulation algorithm, the inverse Gaussian distribution, and the sufficient conditional monte carlo method. Preprint Statistics No. 4/2005, Norwegian University of Science and Technology, Trondheim, Norway.

Thompson, S. K. (1992). *Sampling*. Wiley, New York.

Tuggle, R. M. (1982). Assessment of occupational exposure using one-sided tolerance limits. *American Industrial Hygiene Association Journal*, 43, 338–346.

Wald, A. and Wolfowitz, J. (1946). Tolerance limits for a normal distribution. *Annals of the Mathematical Statistics*, 17, 208–215.

Wang, L. J. and Wang, X. G. (1998). Diameter and strength distributions of merino wool in early stage processing. *Textile Research Journal*, 68, 87–93.

Wani, J. K. (1975). Clopper-Pearson system of confidence intervals for the logarithmic distributions. *Biometrics,* 31, 771–775.

Wiens, B. L. (1999). When log-normal and gamma models give different results: A case study. *American Statistician*, 53, 89–93.

Wilson, E. B. and Hilferty, M. M. (1931). The distribution of chi-squares. *Proceedings of the National Academy of Sciences*, 17, 684–688.

Winterton, S. S., Smy, T. J. and Tarr, N. G. (1992). On the source of scatter in contact resistance data. *Journal of Electronic Materials*, 21, 917–921.

Williamson, E. and Bretherton, M. H. (1964). Tables of logarithmic series distribution. *Annals of Mathematical Statistics*, 35, 284–297.

Xu, Y. L. (1995) Model and full-scale comparison of fatigue-related characteristics of wind pressures on the Texas Tech building. *Journal of Wind Engineering and Industrial Aerodynamics*, 58, 147–173.

Yang, Y., Allen, J. C. and Knapp, J. L. and Stansly, P. A. (1995). Frequency-distribution of citrus rust mite (acari, eriophyidae) damage on fruit in hamlin orange trees. *Environmental Entomology*, 24, 1018–1023.

Zimmerman, D. W. (2004). Conditional probabilities of rejecting H_0 by pooled and separate-variances t tests given heterogeneity of sample variances. *Communications in Statistics–Simulation and Computation*, 33, 69-81.

Zobeck, T. M., Gill, T. E. and Popham, T. W. (1999). A two-parameter Weibull function to describe airborne dust particle size distributions. *Earth Surface Processes and Landforms*, 24, 943–955.

Index